CORN MILLING

Martin Watts

SHIRE PUBLICATIONS

Published in Great Britain in 2008 by Shire Publications Ltd, Midland House, West Way, Botley, Oxford OX2 0PH, United Kingdom.
443 Park Avenue South, New York, NY 10016, USA.
E-mail: shire@shirebooks.co.uk   www.shirebooks.co.uk

© 1983 and 2008 Martin Watts. First published 1983; reprinted 1987; reprinted and redesigned 1998. Second edition 2008.

All rights reserved. Apart from any fair dealing for the purpose of private study, research, criticism or review, as permitted under the Copyright, Designs and Patents Act, 1988, no part of this publication may be reproduced, stored in a retrieval system, or transmitted in any form or by any means, electronic, electrical, chemical, mechanical, optical, photocopying, recording or otherwise, without the prior written permission of the copyright owner. Enquiries should be addressed to the Publishers.

Every attempt has been made by the Publishers to secure the appropriate permissions for materials reproduced in this book. If there has been any oversight we will be happy to rectify the situation and a written submission should be made to the Publishers.

A CIP catalogue record for this book is available from the British Library.

Shire Library no. 98   •   ISBN-13: 978 0 7478 0671 4

Martin Watts has asserted his right under the Copyright, Designs and Patents Act, 1988, to be identified as the author of this book.

Designed by Ken Vail Graphic Design, Cambridge, UK, and typeset in Perpetua and Gill Sans.
Printed in Malta by Gutenberg Press Ltd.

08 09 10 11 12   10 9 8 7 6 5 4 3 2 1

COVER IMAGE
The windmill at North Leverton, Nottinghamshire.

TITLE PAGE IMAGE
Grain feeding from the hopper, along the shoe, into the eye of the runner stone, in Park Mill, Burwash, East Sussex.

CONTENTS PAGE IMAGE
The Walzmühle, one of the first successful roller mills, at Budapest, Hungary, from an article in *The Miller*, May 1890, written to celebrate the mill's fiftieth anniversary.

ACKNOWLEDGEMENTS
I am indebted to many friends and colleagues, millers and mill owners, past and present, for their help and enthusiasm and for enriching my interest in mills and milling. In the preparation of this new edition I am grateful to Luke Bonwick, Peter Hill, John Langdon, Alan Stoyel and James Waterfield, who have all provided information and illustrations. Simon Hudson and the committee of the Mills Section of the Society for the Protection of Ancient Buildings have been supportive as always, as have members of the Traditional Corn Millers' Guild. I am particularly grateful to Sue, my better half, for her encouragement and support; she has read through drafts and made many helpful suggestions, provided useful information from her own study of querns, and searched for illustrations.

Illustrations are taken from material in the author's own collection, except for the following: Bühler Uzwil, Switzerland, page 37 (bottom right); Tracey Elliot-Reep, page 23 (top right); Jane Field, page 46; E. M. Gardner, SPAB, page 22 (bottom right); Peter Hill, page 10 (top right and left); Hitchin Museum, North Hertfordshire Museums Service, pages 19 (top left), and 22 (top); Leicester Museums, page 50 (bottom); Ken Major, page 48; Rural History Centre, Reading University, pages 5, 29 (top right) 30, and 45 (top left); Alan Stoyel, pages 23 (bottom), and 40 (bottom); James Waterfield, page 43 (bottom).

Shire Publications is supporting the Woodland Trust, the UK's leading woodland conservation charity, by funding the dedication of trees.

CONTENTS

INTRODUCTION	4
TYPES OF GRAIN	6
A HISTORY OF CORN MILLING	9
TRANSPORT AND STORAGE	21
PREPARATION FOR MILLING	25
MILLSTONES	31
ROLLER MILLS	36
SIEVING, DRESSING AND GRADING	39
THE FINISHED PRODUCT	44
MILLERS AND MILLWRIGHTS	47
FURTHER READING	51
MILLS TO VISIT	52
INDEX	64

INTRODUCTION

Corn milling is one of the oldest and most necessary service crafts. The main food grains on which the human race depends for an essential part of its daily diet come from plants belonging to the great family of grasses, which cannot be readily digested until the tough outer shells are broken up and, sometimes, removed. A grain of wheat, for example, consists essentially of three parts: the bran, composed of several outer coverings; the germ, the embryo of the new plant; and the endosperm, the starchy centre, which is made into flour.

A rural water-powered corn mill, with a low-breastshot wheel that drove three pairs of millstones and a clover mill: Arrow Mill, Kingsland, Herefordshire.

INTRODUCTION

Bagging up direct from the millstones in a farm mill at Knettishall, Diss, Norfolk, in the 1930s. To the left is a roller mill of the type often found on farms to prepare animal feed.

This introduction to corn milling looks at the methods developed to break open cereal grains in order to extract flour and is mainly concerned with developments and techniques found in the British Isles. The history of corn milling is exceptional in several ways: the necessity of the process and its continuing importance in the production of our daily bread, and the exciting and individual buildings and machines that were developed and that had a marked effect on contemporary technology and the development of sources of energy. The physical remains of the milling industry in Britain are mostly of the last 250 years, a period of great technological change, but their origins can be traced back to the earliest societies. A study of corn milling needs to place mills – both buildings and machinery – in a broader context, however, and it is necessary to keep in mind that mills were constructed to perform useful work, to relieve those who used them from the repetitive drudgery of the daily grind. The relationship of mills to agriculture, to which crops were grown and processed, and to the product requirements of those who built and used them, are also important considerations.

TYPES OF GRAIN

Above: Wheat (left) and barley.

Below: A field of wheat.

WHEAT, originally a native plant of the Middle East, is cultivated over a wider area than any other cereal, being grown in most temperate regions of the world. There are many varieties of wheat, both hulled (with tough outer husks), including emmer and spelt, and free-threshing or naked forms, such as common or bread wheat. The main food value of wheat is as bread flour, the grain containing starch, protein and vitamin B elements. To the miller and baker its important qualities are strength, colour, flavour and flour-yielding properties. Strength is the capacity of a flour to produce a well-risen loaf, and in general red wheats are stronger than white, although white wheats tend to produce flour of a better colour. Britain produces wheats with good flavour, and modern varieties have good bread-making qualities. Wheat has always tended to be the most valued and thus the most costly grain.

Barley was first cultivated in the Near East and, along with emmer, was a staple cereal of ancient Egypt. It was formerly milled for bread flour but it lacks the high gluten content of wheat that makes dough rise. Its main human food value is now in the products of the brewing and distilling industries, for which

the grain requires malting. A small percentage of the barley crop was processed in mills in upland England and Scotland for pearl or pot barley, the skin of the grain being removed by attrition rather than grinding, the grain then being polished in specially constructed machines. Barley is now also important as an animal feedstuff.

The cultivation of *rye*, which was widespread in Europe during the Middle Ages, declined in Britain from about the end of the seventeenth century, up to which time it had provided nearly half the bread corn of the country. Its high gluten content made it responsive to leavening, although it produced dark heavy bread and was therefore often mixed with wheat, a mixture known as *maslin*, or other grains to make a lighter loaf. Grown under varying conditions, rye is less profitable than other cereals and more susceptible to attack by ergot, a poisonous fungus.

Oats, which first appeared as a cultivated crop in Bronze Age Europe, are the hardiest of all Britain's common cultivated cereals, being resistant to both cold and wet. Oats can be grown on most soils, requiring ample moisture and a minimum of sunshine, and were thus always an important crop in upland Britain. The need to dry and de-husk the grain before grinding had an effect on the form of the mills and machinery used to process it. Like barley, oats have a low gluten content and were therefore used for flat breads and oatcakes, as well as oatmeal and porridge. The main value of both grain and straw is now as an animal feedstuff.

Malt is a grain product used in drinks and feeds as a basis for fermentation and to add flavour and nutrient value. Any cereal grain may be converted to malt by steeping, germination and kilning. After being soaked in water for several days, the grain is drained and spread in long heaps and, as its temperature increases, germination begins. After about ten days or so, when the desired amount of germination has been achieved, the sprouted grain, called green malt, is spread on a kiln floor, where it is dried to prevent further germination. In medieval and post-medieval times millstones were used to grind malted grains, mills sometimes having a pair of stones dedicated to malt milling. Iron rollers, which bruised the grains rather than grinding them into meal, superseded millstones for malt milling during the eighteenth century.

Above: A pearl-barley mill in Longhill Mill, Urquhart, Moray. A single vertical millstone is enclosed in a perforated metal casing, into which batches of grain are fed to remove the outer husks and produce pearl or pot barley.

Rye (left) and oats.

A HISTORY OF CORN MILLING

CORN MILLING originated in south-west Asia, where communities who lived in the valleys of the rivers Jordan and Euphrates some ten thousand years ago first gathered the wild grains of wheat and barley that grew there and then started to cultivate them. After harvesting, the grain was pounded in mortars to remove the tough outer husks and was then reduced to *meal*, unrefined flour, by grinding it between stones. The earliest milling tools were rectangular or square slabs of stone with a shallow hollow into which the grain was placed. It was then broken up by the action of a smaller, rounded, hand-held stone that was rubbed to and fro across it. Using this type of mill, known as the *saddle quern*, was hard, time-consuming work, carried out by women or slaves. At Abu Hureyra, an excavated Neolithic settlement in northern Syria, the bones of women showed evidence of deformities, which archaeologists consider to have been caused by many hours spent kneeling and working saddle querns.

Knowledge of agriculture spread into Egypt and then north and west across Europe. Cereals were first cultivated in Britain in the fourth millennium BC, during the Neolithic period, and ground on saddle querns made of locally available stone. At first the cultivation of grain crops was on a limited scale, perhaps just for use during special events, but in the succeeding Bronze Age and Iron Age periods cultivation intensified and small fields were laid out as cereals became a more important part of everyday diet. The saddle quern remained the principal tool for milling grain for more than three thousand years but during the Iron Age, about 400 BC, the *rotary quern*, which comprised two circular stones, first appeared in Britain. The principle of the rotary quern held throughout the development of milling with stones: a stationary lower stone, or *bedstone*, was placed on the ground or on a stand, and an upper, or *runner*, stone was turned above it, while grain was fed through a central hole, the *eye*, in the top stone. The grinding faces of the stones were shaped so that the grain was broken open near the centre, as it entered through the eye, reduced in the central zone and finely milled near the periphery, the ground meal being distributed to the outside edge of the stones for collection. The grinding faces were often pecked or grooved – *dressed* – to give a better cutting action.

Opposite:
A delightful colour illustration of Simmons & Morten's new roller mill at Westminster Bridge, London. In 1885 the mill was working with eighteen pairs of millstones, which were superseded by this roller plant installed by milling engineers William Dell & Son in 1886. There are fourteen double roller mills at first-floor level, eight scalpers above them, seven purifiers at third-floor level, and twenty dressing machines, both reels and centrifugals, on the top floor. The machinery, which was driven by a 50 horsepower beam engine, was capable of producing eleven sacks (about 1.3 tonnes) of flour an hour.

CORN MILLING

Above: Simply described as 'a native mill for grinding corn', this illustration of a saddle quern appeared in *A Popular Account of Dr Livingstone's Expedition to the Zambesi and Its Tributaries*, published in 1887.

Above: Saddle querns were first used over ten thousand years ago and they are still in use in some parts of the world today. Here an African woman is kneeling behind the lower stone, using her weight to crush the grain by moving the upper stone backwards and forwards over it.

Left: Two women working together made the task of grinding with a rotary quern both more efficient and companionable. A song was often sung to set a working rhythm and keep time. The lower stone of the quern shown in this early-twentieth-century postcard has been made in the form of a trough to catch the ground meal.

Below left: The lower stone of a rotary quern. The upper stone has been carefully removed during use to show the milling zones: whole grains at the centre, partly reduced grains in the middle section, and fine meal towards the periphery of the stone.

Below right: Cross-section through a rotary quern, showing the passage of grain and meal, and the rynd, spindle and handle.

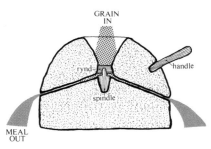

At first thick stones of small diameter with flat or sloping grinding faces were used; these are often referred to as beehive querns from the shape of the upper stone. As the form of rotary querns was developed through use and experience, both stones were gradually made thinner and of greater diameter, and the milling faces eventually became flat. Each development was made to produce more ground corn in less time and with better-applied effort. The rotary quern was both less arduous to use and more productive than the saddle quern; it was quite literally a revolution in milling technology.

Apart from the stones, two other components were essential: a spindle and a handle. The *spindle*, a wooden or iron pivot that projected vertically from the centre of the lower stone, was at first fixed, simply to prevent the upper stone slipping sideways during use. Later the stationary lower stone was perforated through its centre, so that the spindle could pass through and its elevation could be altered to allow regulation of the gap between the lower and upper stones and so control the fineness of the meal, an action known as *lightening* or *tentering*. In order for this to be achieved, the introduction of a fixed wooden or iron bar, the *rynd*, which bridged the eye of the upper stone and enabled it to be hung on top of the spindle, was necessary. A handle, usually of wood, was fitted into a hole in the top or side of the upper stone, or held close to its periphery by a strap or band, and enabled the stone to be turned by hand. The optimum shape and location of the handle must have been discovered by experience through prolonged use, and the diameter of the upper stone of a quern was limited to that which was practical for one, or perhaps two, people to use. The handle was later developed to become a lever for turning the upper stone by man or animal power, allowing an increase in the size and weight of the stone. Eventually the handle was superseded by a power drive applied to the spindle, either from below or from above.

Power-driven millstones were first used in Britain during the Roman period in the first century AD. The classic hourglass-shaped mill turned by a mule or donkey is perhaps the best-known form of Roman animal-powered mill. It is usually referred to as the Pompeiian mill, from the numerous examples found in over thirty-seven milling and baking establishments that have been excavated at Pompeii in Italy. A small number of animal-driven millstones have been identified in Britain, a good example coming from a Roman site excavated in Princes Street, London.

Water power was first used to turn millstones over two thousand years ago. The earliest evidence

An hourglass-type donkey-driven mill, with a baking oven behind, in Regia VI, Insula III, Pompeii, Italy.

The machinery of a horizontal-wheeled corn mill, based on the restored example at the Croft Museum, South Voe, Dunrossness, Shetland. The horizontal waterwheel is turned by water from the open trough striking the flat blades.

suggests that both horizontal and vertical waterwheels were in use about the same time in the lands around the eastern Mediterranean. A horizontal waterwheel is one in which the upper millstone is turned directly by a wheel formed of paddles or scoops radiating from a vertical shaft that is attached to the projecting lower end of the spindle. A vertical waterwheel turns in a vertical plane and thus requires a pair of gears to transfer the motion through ninety degrees, to drive horizontal millstones. The principles of gearing were described and used by the Romans, who introduced the waterwheel and geared mill to Britain. Only a small number of Roman watermills have been identified by archaeologists, from Kent and Hampshire in the south to Hadrian's Wall in the north of England, although large-diameter millstones have been found on nearly two hundred Romano-British sites, which suggests that power-driven mills for grinding grain were far more widespread during the Roman period than was once thought.

The earliest written evidence of mills in England occurs in Anglo-Saxon documents dating from the eighth century. From the ninth century on there is an increasing number of references to mills and their watercourses in charters and place-names. The archaeological evidence for Anglo-Saxon mills is sparse, the two most complete examples having been excavated at Tamworth, Staffordshire, and Ebbsfleet, Kent. Both were horizontal-wheeled mills; the mill at Ebbsfleet, which dated from the late seventh century, had two waterwheels and was a tide mill, using water impounded in a millpond at high tide and released to turn the wheels as the tide ebbed. By the middle

Below: An undershot waterwheel driving a single pair of millstones through a pair of gears, a principle known from Roman times, from William Emerson's *Principles of Mechanics*, 1758.

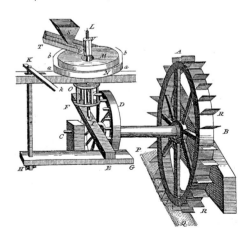

Above: A Roman millstone, Chedworth, Gloucestershire. The dressing pattern shows an awareness of the milling zones, with coarse grinding being carried out near the centre and fine milling towards the periphery.

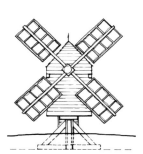  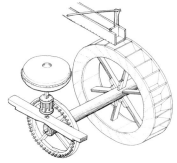

Above left: A medieval post mill.

Above middle: A medieval tower mill.

Above right: A medieval mill, with a single pair of millstones driven by an overshot waterwheel through a pair of gears, a reconstruction based on archaeological and documentary evidence.

of the eleventh century milling was well established in England, with over six thousand mills being recorded in Domesday Book (1086). Whether these were all watermills, and whether they were driven by horizontal or vertical waterwheels or perhaps a combination of both types, remains to be established, although it is usually considered that the mills recorded as manorial assets by the Domesday survey were water-powered corn mills.

Windmills are almost certainly a medieval invention derived from the geared watermill, being first recorded in England during the last quarter of the twelfth century. The earliest type was the post mill, in which the machinery and millstones were located in a timber-built box-like structure that was rotated about the head of a massive vertical post, to face the sails into the wind. The tower mill, with a stone or, later, brick tower that contained the machinery and supported a turning wooden cap that carried the sails and the windshaft driven by them, was also a medieval introduction, the earliest reference occurring before the end of the thirteenth century. Timber-framed tower mills, known as smock mills, appear to have been introduced into England during the sixteenth century.

The nature of the medieval miller's trade, grinding small quantities of grain brought in by each customer, meant that there was no requirement for bulk storage in early mills and little incentive to improve the machinery or output. Most communities would have had a mill within easy reach, often no more than half a day's journey away. From the available evidence, it appears that a medieval mill simply comprised a waterwheel or windmill sails, a pair of gears and a pair of millstones. If an increase in production was required, to satisfy demand, then a second mill was built, sometimes close to the first, and in some areas watermills were supplemented by windmills. By the end of the Middle Ages references to two mills under one roof become more frequent, sometimes with one mill being dedicated to grinding malt. Milling complexes, with several grist mills, a malt mill and perhaps a fulling mill or two, also became more common, particularly where ample water power was available.

CORN MILLING

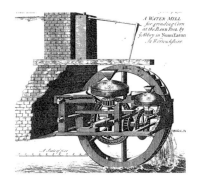

Above: Although the treble mill has been described since the beginning of the seventeenth century, the earliest known depiction of one, with a single waterwheel driving two pairs of millstones, is this engraving of 1723 by Henry Beighton of a mill at Nuneaton, Warwickshire.

Sometime before the end of the sixteenth century the *treble mill* was introduced. This appears to represent the first recorded development of gearing, allowing two pairs of millstones to be driven by a single waterwheel, provided that there was enough power. It is also thought that some post mills were built to drive two pairs of millstones as early as the seventeenth century.

During the eighteenth century milling developed to keep pace with the growth of population and industry. The form of mills had changed only slowly since the Middle Ages, partly because of the limitations and working life of timber, but by the middle of the eighteenth century there was increasing competition for mill sites from a growing number of industries, in particular for producing cloth. Established corn mills often had first rights to watercourses, but their waterwheels and machinery needed improving to make

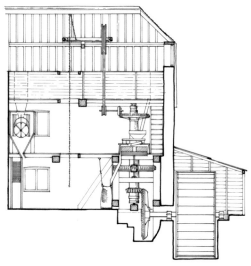

Above: A section through a small rural corn mill of the late eighteenth century, at Coleshill, Oxfordshire. The three-floor layout is typical. Sacks are hoisted up through the mill and the grain is stored in bins in the loft. The middle floor contains the millstones and ancillary machinery, in this case a bolter for dressing flour; and the lower floor, where the miller spends most of his working day, houses the machinery and bagging-off points.

Left: The development of gearing: a medieval mill, with a single pair of millstones; a treble mill, with a waterwheel driving two pairs of stones; and a spurwheel drive arrangement, with two pairs of stones and a crown wheel on the upright shaft, from which ancillary drives are taken. The spurwheel drive became widely adopted in watermills and larger windmills during the eighteenth century.

the best use of available resources. The use of wind power became increasingly important towards the end of the eighteenth century and a number of significant developments were made to the form and technology of both post and tower mills. The most important were the improvements in regulation of sails and millstones and the introduction of iron gearing. The windmill fantail, which was patented by Edmund Lee in 1745, was the first significant invention, although it did not become widespread until nearly the end of the century.

Steam power was first used to drive corn mills in the 1780s, and the building of the Albion Mill, by the Thames at Blackfriars, London, in 1784 greatly influenced the development of milling. Built as a flour factory, the Albion Mill was designed to have twenty sets of millstones driven by two steam engines and was the first mill to be built with all-iron gearing. The mill commenced production in 1786 but was bitterly opposed by millers and others who saw such a massive undertaking as a threat to employment and as having a stranglehold on the price of bread. The mill was destroyed by fire in March 1791 and was not rebuilt, but its short working life marked the beginning of the shift from small country watermills and windmills, often with only two or three sets of stones, to larger merchant mills built close to centres of corn trading and importing. Later, in the nineteenth century, after the repeal of the Corn Laws in 1846 had taken effect, and more particularly from the 1870s, when vast quantities of grain were shipped into British ports from overseas, the milling industry became concentrated at ports and in growing urban centres with good canal, rail or road links.

At about the same time as the building of the Albion Mill, an American millwright, Oliver Evans, wrote *The Young Mill-wright and Miller's Guide*, an important book published in fifteen editions between 1795 and 1860, and translated into French and German. Evans's notable contribution was his design for an automatic mill, in which all the processes of moving the grain and ground stock through the mill to the despatch point were handled by elevators and conveyors. New ideas were often developed only slowly in the milling industry, but within a century the course of milling changed completely.

Below left: The shuttered sails, cap and fantail of the restored Heage windmill, a six-sailed stone tower mill near Belper, Derbyshire.

Below right: The Albion Mill, Blackfriars, London, perhaps the world's first flour factory. The archway provided access from the river Thames for barges laden with grain.

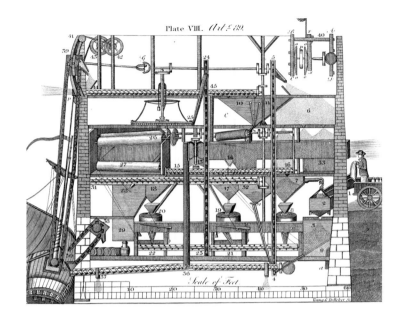

One of Oliver Evans's designs for an automatic mill, from *The Young Mill-wright and Miller's Guide* (1850 edition). Note in particular the use of vertical elevators and horizontal screw conveyors or augers for moving grain and meal.

The method of milling grain between two stones probably reached its peak in the watermills and windmills of the early nineteenth century. The common milling method used in Britain was *low* milling, where grain was reduced to meal in a single run between millstones set close together. The meal was then sifted by mechanical sieves to remove the fine flour. In Europe during the eighteenth century a method known as *high* milling was used. Grain was first broken by being passed between millstones set a little apart so that, although some flour was produced, the starchy part of the grain was reduced to gritty particles termed *semolina*. This was then cleaned, to remove fine flour and bran, before being passed between the stones again, this time set closer together. The ground semolina was again sifted and fine white flour extracted. Using this method, known as *mouture économique*, French millers were able to extract nearly three times more fine white flour than by low milling, thus reducing the amount of low-grade flour produced. Although similar methods were used in some parts of England before the end of the eighteenth century, the softer home-grown wheats were generally ground in a single pass through the millstones.

The use of metal rollers for milling corn had been experimented with as early as the sixteenth century but was generally adopted only for malt milling. Flour mills using iron rolls were tried in Hungary and Switzerland as early as the 1820s but were not particularly successful until Jacob Sulzberger, a Zürich engineer, reconstructed a mill at Frauenfeld, Switzerland, in about 1830. This is claimed to be the world's first successful roller mill, although millstones were still used alongside the rollers. Other mills were built on Sulzberger's principle, using rollers to break the grain and millstones to reduce the semolina. One of the earliest was the Walzmühle at Budapest, built in the late 1830s and rebuilt after a fire in 1851, which had its own millwright's shop and foundry. By 1860 Hungarian millers were using millstones only to clean up low-grade stock and bran, and the revolution from stones to iron rollers was well under way. The new method of making flour, initially referred to as 'Hungarian' or 'New Process' milling, spread to the United States of America and to Britain in the 1860s, where mills using both stones and rollers were built. One of the first in England was that designed by Gustav Buchholz for Fison & Company of Ipswich in 1862. This used a hulling machine to remove the bran, rolls for the first break and two sizes of millstone to produce semolina and reduce it to fine flour. The first 'stoneless' mill is claimed to be that built for W. J. Radford & Sons of Liverpool in 1869, where in 1870 the millstones were replaced by partially grooved rolls. An alternative to chilled iron rolls was the use of porcelain; while claimed to be better than metal, porcelain was not widely adopted because of problems of maintenance and uneven wear. The early attempts to introduce gradual reduction milling into England were not marked with commercial success, however, and it was not until 1878 that the first successful complete roller-mill plant manufacturing flour 'without the use of stones or porcelain rollers' was established in Manchester by Henry Simon. Other Simon mills followed at Tadcaster, North Yorkshire, and at Croydon in 1879, and in 1881 he built the first fully automatic milling plant for F. A. Frost & Sons of Chester.

In February 1878 the National Association of British and Irish Millers (NABIM) was formed at a meeting held in the Corn Exchange Hotel in Mark Lane, London, for 'mutual advancement and protection', considering the 'great changes which are now in progress in the manufacture of flour, and in

Above: Wegmann's improved 'Victoria' porcelain roller-mill unit, from an advertisement in *The Miller*, January 1886.

Right: This advertisement for Seck Brothers' complete roller plants, inserted as a supplement to *The Miller* in January 1887, suggests that their milling system was suitable for use in traditional mills, although in practice the rise of the roller mill resulted in the demise of many watermills and windmills.

the machinery used for that purpose'. Later that year a party of British millers travelled to an international exhibition in Vienna, where at least eight roller-mill machinery manufacturers were represented. In May 1881 an International Exhibition of Milling Machinery was held at the Agricultural Hall, Islington, London. Although it was open for only a few days, the impact on contemporary milling was great. Many millwrights, engineers and mill furnishers, from Britain and other countries, showed complete new milling systems that were soon to become widely adopted in one form or another.

In the late nineteenth and early twentieth centuries milling technology changed at a rate that has parallels with computer technology a century later, where machines and systems were quickly superseded and became out of date. Flour milling became one of the most highly automated industries in Britain. Although many small mills introduced some of the new technology, such as grain cleaners and improved flour dressers, traditional millstone plant was largely superseded by roller mills in the production of fine white flour. Traditional mills that continued in work were often reduced to flour production for small local markets and grinding animal feed. After the formation of the Millers' Mutual Association in 1929, which fixed flour output quotas and raised funds to buy up and close down redundant mills, there was a tendency for large milling companies to acquire smaller firms to take over their quotas. While some small, less efficient mills were modernised, many simply closed down. By the end of the 1930s, when there were about five hundred mills producing flour in Britain, half of which were small independent concerns, three large milling companies controlled about two thirds of the country's flour output.

The modern flour-milling industry, which is still represented by NABIM, is compact and highly efficient, with thirty-one companies operating a total of fifty-nine mills. The two largest companies account for about half of United

A sectional drawing of Mr E. Snowsell's small water-powered roller mill at Nafford, Worcestershire, from *The Miller*, June 1895, shortly after the mill was fitted with new machinery by William Gardner & Son of Gloucester. The space within the original watermill building has been intensively used to accommodate all the grinding and sifting machinery required for the roller-milling process.

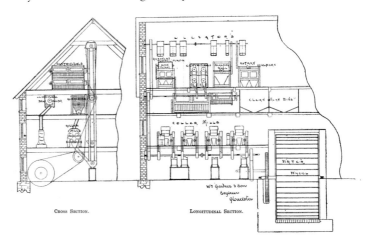

Above left: Ickleford Mill, a watermill on the river Oughton in Hertfordshire, left, was rebuilt in the 1820s and extended in 1892 by Thomas Priest, with the addition of a new steam-powered roller mill using a system installed by the milling engineers Thomas Robinson and Son of Rochdale. The roller mill building still stands, now incorporated in a modern milling complex run by James Bowman and Sons, who bought the mills in 1914.

Above right: Joseph Rank's port mill at Barry, Glamorgan; early twentieth century. The South Wales ports of Barry, Swansea and Cardiff all expanded in the late nineteenth century, with new roller mills built close to the docks to process wheat imported from Russia and North America.

Kingdom flour production. In 2007 over 5.6 million tonnes of wheat were processed, of which up to 83 per cent was home grown.

Traditional milling still continues in some watermills and windmills, many of which were restored to working order during the last quarter of the twentieth century. The demand for stoneground wholemeal, which was an important part of the revival of wholesome, natural foods in the 1970s and 1980s, has evened out, partly because of an increase in the variety of continental and exotic breads that are now available from supermarkets. While many traditional mills grind only on specific days, producing small quantities of wholemeal or animal feed for demonstration, a number are run as small-scale commercial enterprises, using natural power and producing wholemeals and dressed flours from a variety of grains, often concentrating on those that have been grown organically. Traditional mills are well suited to the production of organic meals and flours, as the labour-intensive, craft-based approach to milling, which is essential when using water or wind power, enables small quantities of grain to be selected and milled with care. In 1987 the Traditional Corn Millers' Guild was set up by a group of independent millers, in order to promote stoneground wholemeals, oatmeals and flours to a wider public and to continue the long tradition of milling with stones. In 2008 the Guild had twenty-nine members.

Offley Mill, Staffordshire, a traditional watermill run by the Howell family and still in full production, making stoneground flour.

TRANSPORT AND STORAGE

In the Middle Ages bringing corn to the mill was the responsibility of the customer, and the method used depended on individual status, small quantities being carried manually, while the wealthier tenants and landlords used pack animals. The medieval miller thus received an endless succession of small sacks, the contents of which had to be kept separate and milled within a set time of arrival. As grain can be stored longer than meal or flour, it was brought by the customers when required for use rather than all at once, so that large-scale storage was not needed at the mills. Up to the middle of the eighteenth century the miller's trade was relatively localised, most mills serving an area of only 6 to 7 miles radius. With the expansion of trade during the eighteenth century millers often had their own horses and wagons to collect and deliver corn and flour. Water-borne transport was also important, particularly in areas with poor roads.

To make the handling of sacks of corn easier, facilities such as loading ramps, wagon bays and internal or external hoists that could lift sacks to the top of the mill for storage were installed. The external hoist gable, or *lucam*, became a prominent feature of many watermills, particularly in eastern England. Both windmills and watermills were built taller or larger to accommodate additional machinery and grain storage space, and in newly built mills, particularly in lowland England and in towns, an extra floor was provided so that grain could be emptied into bulk storage bins and fed through spouts to millstones or machines as required. The use of elevators to raise loose grain and move it through the mill was first illustrated by the American millwright Oliver Evans in the 1790s, but elevators were not widely used in Britain in any but the largest flour mills until late in the nineteenth century. In oatmeal mills, however, elevators appear to have been used earlier because the process of separating the husk and cleaning the grain required more handling and machinery than milling wheat flour did. The use of sack barrows, to ease the handling of sacks within the mill, appears to date from the middle of the eighteenth century. Sack handling requires more manpower than elevating or conveying, with one person to hang the sacks on the bottom of

Opposite: A horse-drawn wagon and a Yorkshire steam wagon outside Osney Mill, Oxford, in about 1905. The large lucam projecting from the roof of the mill houses the external hoist for lifting sacks of grain direct to the bin floor for storage under the roof.

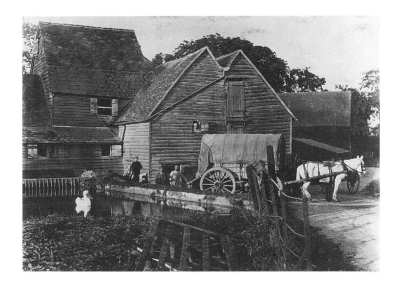

Hyde Mill, on the river Hiz at Ickleford, Hertfordshire; this photograph of c.1900 shows the watermill, left, which was thought to date back to the fifteenth century, with a miller's wagon and horse waiting patiently outside.

Above: This sixteenth-century bench-end in St Bartholomew's Church, Lyng, Somerset, shows a miller returning to his mill with a sack of corn on his shoulder.

the chain and another in the loft to take them off and move the sacks into storage or empty the grain into bins. In modern mills grain is stored in silos, often separate from the mill itself and near to a bulk intake point. The business of manhandling sacks has largely been replaced by automation, the grain being conveyed mechanically or pneumatically through the mill as required.

Right: Delivery of corn and collection of flour and meal by sailing barge and horse-drawn wagons at Brantham Mill, a water- and steam-powered mill on the river Stour in Suffolk, photographed in the early twentieth century.

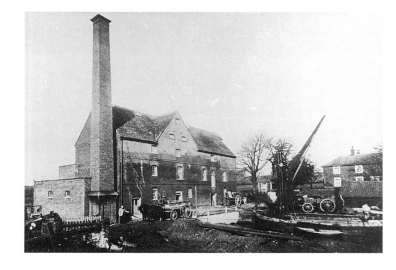

Top left: Carrying to the mill: a sixteenth-century manuscript illustration from Richard Bennett and John Elton's *History of Corn Milling*, 1898.

Top right: Moving and emptying sacks of wheat into the bins in the loft of Crowdy Mill, Devon, in the 1980s.

Middle: The loft of Lapford Mill, Devon, showing the deep bins on both sides of the central walkway, with the sack traps, centre, through which the bags of grain are raised by the sack-hoist mechanism to the right.

Above: The ground floor of a Scottish oatmeal mill, Mill of Mundurno, Aberdeenshire, with elevators, right, for lifting grain and stock up through the mill and a reciprocating sieve and fan in front of the boarded hurst frame, which encloses the machinery and supports the millstones, to the left.

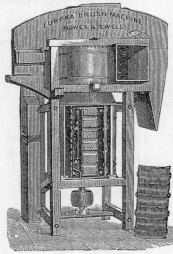

Howes & Ewell's COMPLETE SYSTEM OF EUREKA WheatCleaning Machinery.

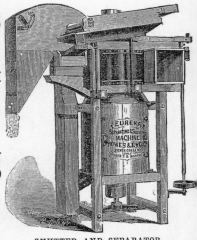

BRUSH MACHINE.

SMUTTER AND SEPARATOR.

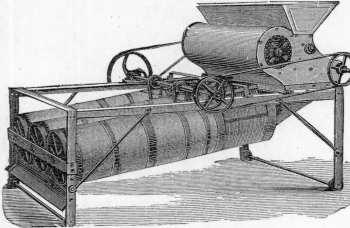

CYLINDERS of All kinds.

For the Complete Separation of all kinds of Seeds from any Wheat.

EUREKA ZIGZAG SEPARATOR.   EUREKA MAGNETIC SEPARATOR.

HOWES & EWELL, 16, Mark Lane, LONDON, E.C.

PREPARATION FOR MILLING

Before the actual milling or reduction of grain takes place certain preparations are usual to ensure the cleanness and quality of the product. In medieval times grain brought to the mill must sometimes have been of indifferent quality and was often a mixture of more than one type. Rudimentary cleaning took place during threshing and winnowing, and coarse sieving removed any large foreign matter that would have been injurious to the stones and the product. From the late eighteenth century grain could be threshed mechanically, and threshing machines produced a cleaned and graded sample. Mechanical cleaning in mills seems to date from about the same time as, and to have developed in parallel with, the increased use of flour-dressing machinery, as the colour of flour is affected by the cleanness of the corn.

Various forms of corn cleaners were developed, the simplest being flat or slightly inclined sieves, shaken either to and fro or with a circular motion while the grain was passed over them. The use of air to blow off dust, dirt, straw and any light particles was important from the earliest time of grain cultivation, and a fan is usual on any machine used for cleaning corn. The *rotary cleaner*, similar to a flour dresser, became common in mills during the nineteenth century; it consisted of a single grade of wire mesh formed around a circular or polygonal drum into which grain was fed to be forced against the mesh by rotating brushes or beaters. A similar machine, in which the cylindrical sieve is usually vertical rather than horizontal or inclined, is the *smutter*, a form of grain scourer used to remove spores of smut, a fungus that attacks wheat. Many of the smutters that survive in British mills are of American origin, introduced from the 1870s, probably as a result of using imported grain and the adoption of gradual-reduction milling.

The mechanical removal of 'wild' corn and weed seeds was also successfully achieved during the nineteenth century with the use of *cockle cylinders* or *trieurs*, which were developed in France. The grain is fed into an inclined drum, which, as it rotates, separates unwanted seeds by size and shape, firstly by fine sieving and then, in the lower part of the cylinder, by a

Opposite: An 1885 advertisement from *The Miller* for a range of grain-cleaning equipment made in the United States of America and imported and sold in London by Howes & Ewell. Like many other mill furnishers, their London address was in Mark Lane, close to the Corn

A rotary corn cleaner in White Mill, Shapwick, Dorset. The cylinder would have been fully enclosed when in use.

rotating indented drum. The shorter, rounder grains are picked up by indents and deposited in a trough. To remove barley or oats from wheat, for example, the wheat grains would be lifted by the indents and led away separately. While cockle cylinders were important for cleaning wheat of all seeds that would damage the flour colour and, like so many of the specialised cleaning and dressing machines, were developed for white-flour production, they are also found in oatmeal mills.

Some of the most injurious matter in grain is pieces of stone and metal, which would damage the milling surfaces and also greatly reduce the life of sieve covers. Magnets are usually introduced during cleaning to pick up pieces of metal, and *rubble reels* or dry stoners are used to separate earth and stones. Reels work on the sieving principle, while *dry stoners* use the difference in specific gravity between the grain and the impurities. The dirty grain is run over a vibrating table, so angled that light grain makes its way in one direction while the heavier matter is vibrated out, to be discharged separately.

A vertical drum smutter in Carew tide mill, Pembrokeshire.

A late-nineteenth-century development was the washing of wheat as part of the cleaning process. This probably began in the large port mills receiving grain from Russia and India and served to clean off surface dirt and to wash out larger impurities. After washing, the grain was dried centrifugally or by being passed through drying machines to restore it to the correct moisture content for milling. Washing is no longer carried out , because of the risk of contamination of watercourses.

One of the most important pre-milling operations is to dry any damp grain so that it can be stored and milled without clogging the milling surfaces or sieves. Force-drying was originally carried out on a domestic scale, although grain-drying kilns associated with corn mills are known from the mid seventeenth century. Kilns were once commonly found beside

PREPARATION FOR MILLING

mills in upland Britain, where high rainfall often caused a late, damp harvest and encouraged the cultivation of oats. Kilns had two main functions: to lower the moisture content of any grain to between 12 and 16 per cent for milling, and to lower the moisture content of oats to such a point that the husks became brittle and could be readily shelled or split off, an important stage in the production of oatmeal.

Some early kilns were circular in form, but the more common survivals are square or rectangular and either freestanding or integral with the mill building. If they were detached, separate grain storage was sometimes necessary. Kilns are usually of two storeys, the lower containing the furnace and the upper the floor on which the grain was spread for drying, which is usually open to the roof. A revolving cowl, ventilator or smoke hole was

Top left: Trieur cylinders, capable of being worked either by hand or by power, from an advertisement of S. Howes of Mark Lane, London, in *The Miller*, May 1889. Working in pairs ensured thorough cleaning and, as with most early milling machines, the action depends much on gravity, feeding at the top and discharging at the bottom.

Top right: A single cockle cylinder in the oatmeal mill at Upper Kennerty, Aberdeenshire. This complete water-powered oatmeal and barley mill was regrettably destroyed by fire in May 2006.

Above: Vibrating vacuum dry stoners at Timms Mill, Goole, East Yorkshire, in the 1980s. These machines are used to separate stones from wheat by their different specific gravity.

Right: A Scottish oatmeal mill with adjoining kiln, Mill of Mundurno, Aberdeenshire. Note the revolving cowl with the pig wind-vane.

Below left: A kiln furnace fired with anthracite at Montgarrie Mill, a water-powered oatmeal mill in Aberdeenshire.

Below right: Kiln-floor construction, with wrought-iron joists and perforated clay tiles, at Bunbury Mill, Cheshire. There are several patterns of clay tile, but most are standardised at 12 inches square.

usually provided in the roof to allow fumes to escape.

The fuel used for firing kilns varied with region and period: oat husks and chaff were formerly used in northern England and Scotland, and peat in the Northern and Western Isles; wood was common in Wales, and later fuels included coke and anthracite. In Wales the supply of fuel was sometimes the responsibility of the customer, who carried it to the mill along with his grain.

Floor construction also varied greatly, the main elements being a fireproof support system, such as slate or stone joists or cast and wrought ironwork, carrying a floor surface of perforated clay tiles, metal plates or sheet, which would retain heat. Metal plates, the dominant floor material used in Scotland, were considered to retain heat better than clay tiles. Grain was spread to a depth of 4 or 5 inches over the floor, so the quantity that could be dried at any one time depended on the original moisture content, the heat of the fire, the size and construction

PREPARATION FOR MILLING

of the kiln and the wishes of the miller and his customer. When dried, the grain was bagged or led through spouts to the mill for shelling or grinding.

In the mid nineteenth century a new form of kiln was introduced, which comprised a large vertical metal cylinder in which the oats were moved automatically from top to bottom, being dried by hot air, the temperature of which was thermostatically controlled. Kiln drying has now almost died out, and kilns that survived into the twentieth century were used chiefly for the production of oatmeal. Firing took place when required and could become a social occasion, the heat of the furnace providing a warm meeting place and somewhere for the villagers to cook a few extras. In the modern industry drying is part of an automated process, the cleaned and graded wheat passing through specialised driers on its way to be ground.

Top left: The kiln floor or 'head' at Muncaster Mill, Cumbria.

Above: Turning grain in the kiln at Viver Mill, Hincaster, Cumbria, in 1941. Turning the grain to ensure even drying must have been an unpleasant task in the smoke and fumes.

Left: A Walworth kiln, introduced for drying oats in the mid nineteenth century.

29

MILLSTONES

MILLSTONES used in British mills are from quarries either in Britain or on the continent of Europe and are of two main types, natural or manufactured. The use of more than one type of stone in a single mill was common, being related both to what was being milled and to fashion.

The most widespread British stones are the sedimentary gritstones, particularly Millstone Grit *Peaks* or *greys*. The main source of these was the Peak District of north Derbyshire, and, considering the remote inland location of the quarries, the stones are found over a remarkably wide area. Latterly their main use was for grinding animal feeds, for their gritty texture gives a soft meal and cuts up the bran too finely for clean dressing out. Welsh stones, of sandstone conglomerate, were quarried in the Wye valley and on Anglesey and similarly used, although more local in distribution. Igneous rocks were also used, in particular granite in Devon and Cornwall, and many small quarries were worked on a local scale in attempts to find stone of character and quality comparable to that imported from Europe, particularly in times of recession or war. British stones are usually monolithic, although some examples built up of shaped segments bound together are known, as for example with millstones from Mow Cop, on the Staffordshire–Cheshire border.

Millstones were imported from Europe as early as Roman times, the main sources being quarries in the Mayen and Niedermendig districts of Germany and, later, those of La Ferté-sous-Jouarre and Épernon in France. German millstones, of a dark-coloured basalt and known as *blacks*, *blues* or *Cullins* (after Cologne, a major export centre), were formerly quite common, particularly in northern and eastern England. They were said to wear quickly and give off a dark powder that, while it would not have been noticed in dark rye bread, would certainly have discoloured wheat flour. They were also used for grinding malt.

The French quarries produced what was generally considered the finest stone for wheat milling and white-flour production. *French* stones or *burrs* were usually built up of shaped and faced segments of a quartz-like stone, a tertiary chert found in the Paris basin. Although solid stones are found, in

Opposite:
A finished composition millstone with a conventional land and furrow dressing, rolled out for transit at the Silex Works, Isle of Dogs, London, in 1939.

31

Above: Peak millstones abandoned alongside the track from the quarry at Bole Hill, in the Peak District of Derbyshire.

Above right: An advertisement for Peak millstones from *The Miller*, 1888, indicating some of the uses to which these stones were put.

Below left: How millstones work: A is the swallow, to draw in the grain, B the reducing zone and C the grinding zone. Both milling faces are dressed identically and divided into harps or quarters, each with four furrows cut between the lands; r is the rynd or bridge, by which the runner stone is hung on the mace, m, at the head of the spindle, and turned by it.

Britain stones were generally manufactured from smaller blocks of stone that were often imported as ballast. The stone was graded, shaped, fitted together with cement and a plaster backing, and finally bound with iron hoops around the circumference. In the nineteenth century the manufacture of French stones became a major industry concentrated at ports and milling centres, and their ubiquitous use in English mills underlines the contemporary demand for white flour for, when well dressed, French stones mill crisply and leave the bran in large flakes for easier dressing out. The cost of a French stone was three or four times that of a local one and, from the Middle Ages onwards, the cost of millstones, including their transport, could account for a large proportion of the building and maintenance costs of a mill.

In the later nineteenth century composition millstones were developed, being a manufactured bond of cement with a granulated material, such as burr chips or emery, to form the milling surface. The irregular faces required less frequent dressing than French stones, and they were less expensive,

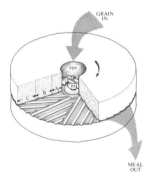

Above right: A granite bedstone in Trewey Mill, Zennor, Cornwall. The iron fitting in the centre is the rynd or bridge, which is slotted over the head of the millstone spindle to carry and turn the runner stone.

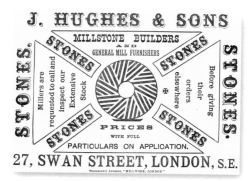
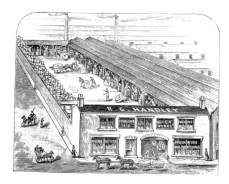

Top left: A Cullin millstone with closely dressed furrows and lands, leaning against the roundhouse wall at Stanton post mill, Suffolk. The chases cut into the stone close to the eye indicate that it was located and turned by a four-armed stiff rynd.

Top right: A French burr bedstone in Crowdy Mill, Harberton, Devon. The joints between the burr blocks and the furrow and land dressing can be clearly seen.

Above left: An advertisement for one of the London-based millstone builders, from *The Miller*, 1888.

Above right: The millstone works of R. G. Handley in Birmingham, from *The Miller*, April 1889. Millstone makers and dealers became established in several of the larger towns, such as Birmingham, Gloucester, Hull, Leeds and Liverpool, during the second half of the nineteenth century.

but they were not developed until the use of roller mills was already under way, and their chief use was for milling provender or animal feed, in both horizontal and vertical stone mills.

Much of the skill of the traditional miller and millwright was in the dressing and setting up of millstones. In Roman times the grinding faces of millstones were sometimes cut with a pattern of *furrows*, particularly towards the periphery of the stones, where fine grinding took place. The faces of Anglo-Saxon millstones appear to have been simply pecked to produce a random series of cutting edges. Furrow dressing reappears on millstones after the Norman conquest, although the few examples that have been identified are quite crudely laid out, the furrows being cut by a sharp pick, which was also used to peck the *lands*, the flat areas between the furrows on which the

Top left: A composition millstone built into a wall outside Shaw Mill, Berkshire. Note the lighter-coloured grinding medium of the swallow, around the eye, and the curved or sickle-shaped furrows. These millstones were usually used for milling provender or animal feed.

Top middle: A detail of millstone dressing, showing narrow furrows cut with a pick, and pecking on the lands to produce cutting edges, on a Peak millstone that probably dates from the sixteenth century, at Thornton Abbey, Lincolnshire.

Top right: A French burr bedstone leaning against the kiln wall at Heron Mill, Beetham, Cumbria. The widely spaced furrows indicate that this was one of a pair used for oat shelling.

Above: A millstone dresser at work, here seen stitching the lands of a Peak runner stone – an evocative photograph taken in the 1920s by W. H. Palmer, from William Coles Finch's *Life in Rural England*.

Above: Precision dressing on the lands of a French burr stone in Houghton Mill, Cambridgeshire. The fineness and regularity of the stitching suggest that it was probably cut by a millstone-dressing machine.

Above: The cover plate of a balance box set into the plaster back of a French millstone in Houghton Mill. Lead can also be seen between the balance box and the periphery of the stone, suggesting that it must have been a problem to balance the stone precisely.

reduction of the grain particles took place. By the eighteenth century more precise patterns were being used, particularly for flour milling. Millstones used for grinding malt generally have a closer concentration of furrows, as wide lands were not required for milling a granular product. Stones used for shelling oats usually have few or no furrows. When working, they are set a little apart in order to nip the husks off the oat grains, after which the shelled oat kernels, called *groats*, are separated by sieving before being ground on a second pair of millstones. The ground meal is then sifted to produce various grades of oatmeal.

The amount and frequency of dressing depends on the use of the stones, those for flour milling being generally better attended. The craft of millstone dressing is a mixture of skill and hard work, the correct setting out of the furrows and lands being followed by the labour of facing, furrowing and stitching or cracking. *Facing* is the process of ensuring that the milling faces are plane; *furrowing* entails cutting the furrows to the correct depth and profile, and *stitching* is lining the lands with parallel cracks precisely enough to obtain maximum grinding performance from the stones. The mill bills, chisels and picks used for stone dressing may seem crude tools, but in experienced hands they can produce fine work. Finally the stones must be set up to run true, and the runner stone must be correctly balanced and hung on the spindle so that it does not dip or wobble when working, which would cause the milling faces to touch and reduce both the life of the stones and the quality of the meal. French stones and manufactured composition stones usually have pockets let into their backs in which weights are put to balance them accurately.

During the nineteenth century many experiments were made by millers trying to improve both the grinding performance and the output of millstones, particularly as new ideas about grain reduction and flour production were becoming a practical reality with the development of roller milling. As well as refinements in the design and execution of millstone dressing, including the use of millstone-dressing machines, a ventilation system for millstones was also introduced. By drawing a controlled current of air through the millstones, it was found that they worked better when cooled, and output was increased.

Below left: An overdriven French burr millstone in Houghton Mill. The balance boxes and maker's nameplate can be seen set into the plaster back of the runner stone. The timber ducting behind the stones is the remains of a millstone-ventilation system introduced into the mill by miller Potto Brown in 1846.

Below right: The stone floor of Trafford Mill, Cheshire, with three pairs of French stones complete with their millstone furniture and spoutwork. Local and regional variations in the design of millstone furniture are just one of the aspects that make the study of traditional mills so absorbing, and historically accurate preservation so important.

ROLLER MILLS

THE PRINCIPLE of roller milling is to reduce the grain gradually between pairs of rolls, subjecting the ground material to one or more separations between each reduction. The process is gradual in order to control particle size precisely and produce as much white flour as possible.

Originally there were three main groups of roller mills, break rolls, scratch rolls and reduction rolls, each working on the principle of passing grain or partly reduced stock between two rollers that move at different speeds. The slower roll holds the stock while the faster roll cuts. *Break rolls* are fluted to give a slight scissor-like action that assists in tearing open the grain and scraping the endosperm from the bran skin. *Scratch rolls* have a similar appearance but with finer flutes, their function being to reduce the size of semolina granules and to remove any bran that still adheres. *Reduction rolls* are smooth and are used finally to reduce the granular stock and convert it to flour; their action also makes it possible to separate any small pieces of bran that remain. In most modern systems only break and reduction rolls are used.

At each stage of the reduction process some flour is made, if only by attrition, so each pass through rolls the stock is subjected to some form of separation: scalping, grading, dusting, purifying or flour dressing.

Right: A heavy-pattern four-roller mill, from Henry Simon's 1892 catalogue.

Far right: A pair of porcelain rolls displayed on a stand in Houghton Mill, Cambridgeshire.

ROLLER MILLS

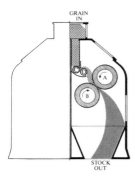

Top left: A section through a diagonal four-roller mill. Each unit is divided into two halves, which can be used on separate runs. A is the fast roll, B the slow, C and D the back and front feed rolls necessary to ensure an even stream of grain or stock on to the main rolls.

Top right: The roller floor of Caudwell's Mill, Rowsley, Derbyshire, a water-powered mill run by the Caudwell family from 1874 to 1978. Some of the roller-mill units were installed by Briddon & Fowler of Manchester in 1905, and in 1914 the machinery was updated, and a new water turbine installed, by Amme, Giesecke & Konegen, a German company, whose millwrights were interned for the duration of the First World War.

Scalping is a coarse separation following each break, to separate the endosperm; *grading* is the separation of single stock into two or more groups, according to particle size; *dusting* is the removal of flour and other material from the stocks before they are purified; *purifying* is the separation of fine bran particles and pieces of endosperm with bran adhering from the pure endosperm before it is ground to flour; and *flour dressing* is the sifting out of oversize particles from the finished flour. In each process separations are made by particle size over sieves, small particles passing through while larger particles remain for further treatment. These separations have led to the development of several specialised machines, some being preferred to others in certain milling systems. With the exception of the *purifier*, which uses flat sieves and air currents to perform separations, these machines have their origins in the early dressing machinery developed for use in stone mills.

Above: The roller floor of Timms Mill at Goole in the 1980s, with double diagonal roller-mill units by the Swiss milling engineers Bühler and Robinsons of Rochdale.

Above: Inside a twenty-first-century mill: Bühler roller mills at San Miguel, Salvador, Central America.

37

SIEVING, DRESSING AND GRADING

THE product of millstones is normally termed *meal* until some further processing has taken place. For bread flour, some bran is usually removed by sifting to give a lighter texture and better colour, but much depends on demand, fashion and economics. White flour and the bread made from it were always regarded as something of a delicacy; the Roman historian Pliny observed that the excellence of the finest kinds of bread depended on the fineness of the sieve cover. White flour was, however, costly to produce, for up to the sixteenth century most sifting was done manually. The demand for white flour in medieval times was probably influenced by religion, white bread being symbolic of purity.

It must have been noticed in toll mills that when meal fell from the millstones into the meal bins or arks natural gravity separation took place. As the meal formed a pyramid under the end of the spout, the coarse bran would fall down the outside, leaving the finer flour in the centre. Early flour dressers were sometimes fed direct from the meal spouts and shaken in some way by the same drive as that to the stones. Mechanical dressing machinery follows one of two basic principles: flat or inclined sieves driven with an oscillating or gyratory motion, imitating the hand-worked sieve; and revolving cylindrical or polygonal sieves, with or without external beaters or brushes. The first type was probably the earliest used in windmills and watermills, for example the *jog-scry* or *jumper*, a tiered sieve often used for sifting shelled or partly ground corn before a second run through the stones. In 1770 John Milne, a Manchester wire worker, patented a sifting machine with three sieves one above the other, all actuated by a central shaft. In principle this anticipated the *plan sifter*, which uses a large number of flat sieves and is widely used in gradual-reduction milling systems. Flat sieves have a higher capacity and require less driving power than rotary machines.

Of rotating dressing machines, three types were traditionally used: the bolter, the wire machine and the reel. The principle of the *bolter* was that the ground meal was fed into a cylindrical cloth sleeve, against which it was held by centrifugal force while the cloth was rotated and knocked against fixed

Opposite: Reciprocating sieves or jog-scries positioned above the millstones in Heron Corn Mill, Beetham, Cumbria. The raised hurst frame, known locally as a 'lowder', carries four pairs of millstones. The two pairs on the left side were used as part of the oatmeal process, with ground stock being lifted from their meal spouts by elevators, to be sifted by the jog-scries.

CORN MILLING

Gravity separation of meal as it falls from the spout into the meal ark at Worsbrough Mill, South Yorkshire. Fine flour is concentrated in the centre, while the coarser particles and bran fall down the outside.

bars so that the fine flour was driven through. Bakers originally worked bolters by hand, horse or sometimes wind power, and bolting houses were built alongside some mills in the sixteenth and seventeenth centuries. Later in the seventeenth century bolting machines were introduced into mills and driven by gearing or belts from the main machinery.

In 1765 John Milne took out a patent for a machine with an inclined fixed cylindrical drum covered with different grades of wire cloth, the finest mesh being at the top and the coarsest at the bottom. Meal fed into the top of the cylinder was brushed against the cloth, so that fine flour was separated

A small rural corn mill, Bossava Mill, Paul, in west Cornwall, photographed in the 1890s. The spout from the millstones directs the meal into a small sieve located in the ark or meal bin. Note the single pair of millstones driven directly from the pitwheel, in the medieval manner, and the west Cornish design of the millstone case.

SIEVING, DRESSING AND GRADING

Above: A bolter in Arrow Mill, Kingsland, Herefordshire. The bolting cloth has long gone, but the timbers that supported it and the fixed beater bars can be seen, as can the drive pulley at the top right.

Top left: Plan sifters in Caudwell's Mill, Derbyshire. Each sifter contains a number of flat sieves through which ground stock is passed to separate fine flour from coarser particles, which are then subjected to further grinding.

Top right: A small plan sifter for removing coarse middlings and bran in Elsecar Mill, South Yorkshire, where stoneground flour was produced by Allied Mills Ltd in the 1980s.

A wire machine, showing the drive and hopper feed at the upper end of the cylinder, in Felin Newydd, Crug-y-Bar, Carmarthenshire.

at the top, increasing in coarseness down to bran at the bottom. Advantages of the *wire machine* over the bolter were that several grades could be separated in one 'through' and its capacity was high because of the speed of rotation of the brushes. The wire cloth also lasted longer than the cloth sleeve of the bolter. However, flour colour was lowered because of the speed, and more driving power was required.

Reels were in use in France in the eighteenth century but apparently were not adopted in English mills until the mid nineteenth century. They were generally polygonal in form, usually six-sided, and could be up to 28 feet in length. They normally had finely woven silk covers for flour dressing and brass wire cloths for sifting coarser stock. An advantage over both bolters and wire machines was their slow speed and gentle action, which did not break up the bran particles and thus enabled a high-quality good-colour flour to be extracted. They were often used in town mills and alongside roller plant, but some large machines were installed in small rural mills. A disadvantage was that only about one third of the sieving area was effective at a time,

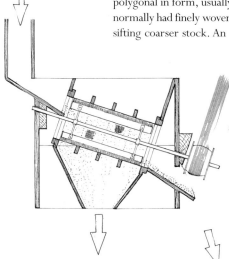

Diagram showing the working of a small wire machine, for separating fine flour from bran. Meal fed into the cylinder at the top left is brushed against the wire screen, the brushes forcing the fine flour through the mesh, with the bran and coarse particles falling out at the bottom right. Larger machines usually had two or three dividers in the hopper section beneath, to separate different grades of fine flour.

and they were often installed in pairs so that the 'over-tails' from the first would pass through the second for further separation. Sifted flour was usually conveyed from the bottom of the reel case by a horizontal screw conveyor. The introduction of rapidly revolving internal beaters to throw the stock against the sieve surface was an improvement that came with gradual reduction milling and led to various forms of *centrifugal* separator.

A partly dismantled Simon reel, with the silk cloth hanging to the right, in Great Alne Mill, Warwickshire. The silk would be stretched over the hexagonal frame and the sifted flour conveyed from the bottom of the case by a screw conveyor.

A Hopkinson centrifugal separator in use for making white flour in the five-sailed Maud Foster Mill, Boston, Lincolnshire.

THE FINISHED PRODUCT

In former days the customers' meal from the arks or millstone spouts was bagged in the sacks in which the grain had arrived, allowing for a reduction in volume or weight by the subtraction of the miller's toll. Grain measurement was by volume, the bushel measure being widely used. Later, steelyards and beam scales were used to weigh grain on arrival and flour before departure, but these have been superseded by platform scales, and a large modern mill will have a weighbridge.

In traditional mills a space was often set aside by the main door for an office, which might have been a small room or simply a wooden desk fixed against a wall, and there records of milling accounts and charges were made. A sack of flour became standardised at 20 stone or 280 pounds in the nineteenth century, and the output of a mill is still often expressed in sacks per hour or per week rather than in tons, imperial or metric.

The miller's workplace: meal arks and millstone controls on the ground floor of Manor Mill, a small rural watermill with two pairs of millstones at Branscombe, Devon.

THE FINISHED PRODUCT

Left: Bagging up and weighing wheat for seed. This photograph, taken in 1948, shows a traditional type of volume measure and a deadweight weighing machine in use, with a winnower in the background.

Traditional hessian sacks and cotton flour bags are now almost obsolete, grain and flour being delivered in bulk or packed in paper sacks and loaded on pallets by forklift trucks, rather than being manhandled. The finished products are also mechanically packed and handled.

Much of the development of milling has been due to the practical experience of many generations of millers and millwrights, all trying to make a product to their satisfaction and that of their customers. In corn milling there are many variables because of the dependence of the industry on agriculture, but the quality of the product depends on three things: the nature and sample of the grain; the state and management of the machinery; and, above all, the immeasurable skill and experience of the miller.

Below left: The miller's desk in a quiet, well-lit corner of a country mill, Arrow Mill, Kingsland, Herefordshire.

Below right: A warning that ownership of bags must be made clear, nailed to a beam in a Welsh country mill.

45

MILLERS AND MILLWRIGHTS

THE name *muilleóir* for a specialised miller occurs in a ninth-century Irish law text. In England a miller named Wine, at Langford in Bedfordshire, is mentioned in a late-tenth-century will in a context that indicates he was a servant rather than a free man. Only six people, none of them named, are referred to as millers in Domesday Book, as widespread as Cheshire and Sussex. As over six thousand mills were recorded by the Domesday survey, it is unlikely that such a small proportion had millers, rather that many mills were worked by manorial servants or tenants of middling status whose names and occupations are unrecorded. The name *mylner*, denoting a maker of meal, does not appear until the thirteenth century, when occupational surnames such as Miller, Milner and Milward developed from names such as 'John the miller'. Robyn and Symkyn, the millers of Geoffrey Chaucer's *Canterbury Tales*, are both larger-than-life caricatures, full of popular imagery that a contemporary audience would have understood and appreciated. Deceit, greed, cunning and lechery were all characteristics associated with medieval millers. A manorial miller undoubtedly did have the opportunity to steal from those whose grain he ground by taking more than his due as payment in kind for the service; he was, after all, in control of the toll dishes and corn measures, and his 'golden thumb' enabled him to judge the quality of the grain brought in for milling. But although manorial and court records indicate that millers were sometimes fined for taking excessive toll, when looked at in the context of the ten thousand or so watermills and windmills that were working in England by about 1300, the percentage of offenders was rather small.

An important influence on building and working mills in medieval times was the milling *soke*, the right or privilege that gave landowners the power to build and work corn mills and that bound many manorial tenants to give their grinding custom to such mills, although there were numerous exceptions. Mills were an important source of manorial income, and water rights, as well as the milling soke, were often jealously guarded by those who owned or benefited from them. The basis of the medieval millers' trade was to grind all the corn consumed on the manor, whatever its origin, and to take a toll by way of

Opposite: Dick Blezzard, a millwright from Preston, Lancashire, and his men standing in front of the five-sailed tower mill at Tuttle Hill, Warwickshire, newly rebuilt after storm damage in 1905. By the early twentieth century traditional millwrights were decreasing in number and often travelling greater distances to work on mills.

47

Bringing corn to a windmill: a sixteenth-century misericord in Bristol Cathedral. This delightful carving illustrates a medieval joke told against millers. A miller is returning to his mill leading his packhorse or donkey, which is carrying a sack of grain. Seeing the animal is tiring, and being still some way from the mill, the miller thoughtfully shoulders the sack, but then mounts the poor beast and rides it back to the mill!

payment. The amount of toll varied with region and period, from one thirteenth part of the grain taken by volume in the north of England to as little as one thirty-second, or even less, in the south and west. In the later Middle Ages the social status and prosperity of millers varied greatly, from lowly paid manorial servants to middle-level peasants, who were sometimes miller-proprietors. Unlike many other craftsmen and tradespeople, English millers were never protected by, or benefited from, guild status. The relatively isolated locations of many mills meant that millers were too scattered geographically to form guilds, unlike the bakers who supplied the growing urban areas.

There were significant changes in land ownership after the dissolution of the monasteries in the late 1530s, together with improvements in the standard of living. Mill owners and millers were now in a position to improve both the performance and the output of their mills, to meet demand as the population rose. Although there were still disputes about the enforcement of the landlords' customary rights connected with the use of mills, which had become well established during the Middle Ages, millers gradually became more competitive and were able to buy grain and sell meal and flour for profit. Between 1640 and 1750 mill monopoly declined significantly in English market towns, although the milling soke persisted in some areas, particularly in the north of England. Grist milling, in which small batches and gleanings were ground for individual customers, was far less profitable than the merchant trade and was generally incompatible with town life. In 1796 an Act for the Better Regulation of Mills, which attempted to replace the toll-in-kind system with a monetary charge, was passed and millers were obliged to put up tables of charges for grinding and dressing meal. By the end of the Napoleonic Wars, the restrictions of soke had virtually disappeared, although at Wakefield, one of the most important grain markets outside London, the milling soke was not finally abolished until 1853. In 1851 there were some 37,000 millers in Britain; some of the larger steam mills employed fifty to a hundred workers, but the majority of small mills, which were scattered throughout the countryside, were family-run, knowledge of the miller's craft being passed from father to son, sometimes through several generations. With the great changes in milling technology that were introduced during the last quarter of the nineteenth century and the subsequent automation of flour milling that followed, it has been estimated that only some fifteen thousand operatives were directly employed in the milling industry in the late 1940s.

To build and maintain the thousands of mills that existed in Britain at the time of the Norman conquest, the millwrights' craft must already have been well established. Probably the earliest reference to a specialised builder of mills comes from early medieval Ireland, where the *sáer muilinn*, who was afforded similar status to other master craftsmen, equivalent to the lowest grade of noble, is referred to in a seventh-century text. Mills and their machinery were built largely of timber and in medieval building accounts it is carpenters who appear to be in charge of constructing mills, aided by blacksmiths, who forged iron fastenings and parts for the machinery, such as spindles and bearings. It is not until the later fourteenth century that the specific title millwright first appears in England.

From the seventeenth century increasing demands were being made to improve the efficiency and output of watermills and windmills. Water-powered sites in particular were being developed for a growing number of trades and industries, such as textile production, metalworking and papermaking, and the role of the millwright grew accordingly. By the early eighteenth century the power of both water and wind was being investigated scientifically and a new breed of engineers, men such as John Smeaton and John Rennie, whose roots were in traditional millwrighting, grew up. The increasing use of cast iron led to improvements in the lasting quality of waterwheels in particular and to a great change in the main mechanical elements of shafting and gearing. By the 1820s iron was in widespread use for engineering components, with urban and rural foundries producing items for mills and agriculture in general.

Frank Mettam, watermiller, and Bill Heathershaw, windmiller, beside the stones in Ollerton Mill, Nottinghamshire, in 1979. The dusty atmosphere of an unrestored working mill and two traditional stone millers talking together represent a situation that has almost completely disappeared since the 1980s.

Right: A bushel measure of rye, with a strike for levelling the top of the grain, illustrating the traditional way in which grain was measured at a mill.

Far right: A rare survival: charges for grinding and dressing painted on the end of the bolter case in Arrow Mill, Kingsland, Herefordshire.

The design and construction of mill machinery became increasingly sophisticated and foundry-made components were often transported considerable distances, to be fitted by local millwrights, who still maintained the craft skills and knowledge to build and maintain both watermills and windmills.

The 1851 census of Great Britain recorded about ten thousand millwrights, ranging from traditional craftsmen who built and serviced watermills and windmills, to those attached to large engineering concerns. From this time mechanical and civil engineering began to supplant the more traditional, vernacular nature of the craft. During the second half of the nineteenth century much milling machinery, particularly for cleaning grain and dressing flour, was also factory-made, sometimes being imported from abroad, and what has been described as the 'brown box' era of mill machinery became firmly established. In the late nineteenth and early twentieth centuries the role of the traditional millwright, as indeed that of the traditional stone miller, became increasingly diminished as the method of making flour by gradual reduction in roller mills took over from the millstone mills that worked by water and wind power. The role of the traditional millwright in the early twenty-first century is largely that of a repairer and conservator of historic buildings and machinery. It is essential that the traditional mills that do survive should be authentically repaired and continue to serve as a potent reminder of the vital relationship between agriculture and food, and the necessity of producing flour for mankind's daily bread.

The millwrighting works of William Cartwright at Wellington, Leicestershire, photographed in 1897. Note the thick French burr millstone and the casting pattern for a large bevel-geared pitwheel. The sheerlegs, left, and jacks indicate something of the heavy handling equipment that nineteenth-century millwrights had at their disposal for building and repairing watermills and windmills.

FURTHER READING

There are many books on watermills and windmills, including some good local and regional studies, and several on the modern milling industry, but few which deal with the milling processes and products, particularly regarding the traditional industry. Most useful are:

Bennett, Richard, and Elton, John. *History of Corn Milling* (four volumes). Simkin Marshall & Co, 1898–1904. The classic and most historically complete account of the industry up to 1900.

Gauldie, Enid. *The Scottish Country Miller 1700–1900.* John Donald, 1981. One of the few publications that concentrate on the social rather than the technical side of milling.

Jones, Glyn. *The Millers. A Story of Technical Endeavour and Industrial Success, 1870–2001.* Carnegie Publishing Ltd, 2001. A detailed study of the rise of the modern milling industry in Britain.

Kick, Friedrich. *Flour Manufacture.* Crosby, Lockwood & Son, 1888.

Kozmin, Peter. *Flour Milling.* George Routledge & Sons, 1917. Like Kick, this is a translation into English but both are valuable contemporary studies of the change from millstones to rollers.

Langdon, John. *Mills in the Medieval Economy, England 1300–1540.* Oxford University Press, 2004. A scholarly and readable account of milling in medieval England.

Lockwood, J. F. *Flour Milling.* Henry Simon Ltd, fourth edition 1962. The modern standard on the subject.

Moritz, L. A. *Grain-mills and Flour in Classical Antiquity.* Clarendon Press, 1958. A well-ordered, thorough and fascinating study.

Scott, J. H. *Development of Grain Milling Machines.* Turret Press, 1972. A fully illustrated description of a wide variety of milling machinery.

Voller, William R. *Modern Flour Milling.* Gloucester, 1897. A valuable study of gradual-reduction milling practice by an author engaged in the practical education of millers.

Watts, Martin. *Water and Wind Power.* Shire Publications, second edition 2005. A chronological account of mills and traditional sources of power in Britain.

Watts, Martin. *The Archaeology of Mills and Milling.* Tempus Publishing Ltd, 2002. An overview of the physical evidence for corn milling to be found in the British Isles.

MILLS TO VISIT

Many museums have querns and other early milling items on display in their archaeology and local history galleries. Some modern mills welcome organised parties and interested individual visitors by prior arrangement. Preserved water-powered roller mills can be visited at Caudwell's Mill, Rowsley, Derbyshire, and Calbourne, Isle of Wight. Many watermills and windmills are regularly open to the public, some with querns and other interactive displays for visitors. A number of traditional mills still work and have flour for sale. Readers are strongly advised to check opening times and dates before making a special journey to a mill.

WATERMILLS

BEDFORDSHIRE

Bromham Mill, Bridge End, Bromham, Bedford MK43 8LP. Telephone: 01234 824330. Website: www.bedford.gov.uk

BUCKINGHAMSHIRE

Ford End Mill, Station Road, Ivinghoe LU7 9EA. Telephone: 01442 825421. Website: www.fordendwatermill.co.uk

Pann Mill, London Road, High Wycombe. Telephone: 01494 472981. Website: www.pannmill.org.uk

CAMBRIDGESHIRE

Hinxton Mill, Mill Lane, Hinxton. Telephone: 01223 243830. Website: www.cpswandlebury.org

Houghton Mill, Houghton, Huntingdon PE28 2AZ. Telephone: 01480 301494. Website: www.nationaltrust.org.uk/houghtonmill

Lode Mill, Anglesey Abbey, Quy Road, Lode, Cambridge CB25 9EJ. Telephone: 01223 811080. Website: www.nationaltrust.org.uk/angleseyabbey

Sacrewell Mill, Sacrewell Farm and Country Centre, Thornhaugh, Peterborough PE8 6HJ. Telephone: 01780 782254. Website: www.sacrewell.org.uk

CHESHIRE

Bunbury Mill, Mill Lane, Bunbury CW6 9PP. Telephone: 01829 261422. Website: www.bunbury-mill.org

Stretton Mill, Mill Lane, Stretton, near Farndon SY14 7JA. Telephone: 01606 41331. Website: www.strettonwatermill.org.uk

CORNWALL

Cotehele Mill, St Dominick, near Saltash PL12 6TA. Telephone: 01579 350606. Website: www.nationaltrust.org.uk

Melinsey Mill, near Veryan, Truro TR2 5TX. Telephone: 01872 501049.

Trewey Mill, Wayside Folk Museum, Zennor, St Ives TR26 3DA.
Telephone: 01736 796945.

CUMBRIA

Acorn Bank Watermill, Temple Sowerby, near Penrith CA10 1SP.
Telephone: 01768 361893. Website: www.nationaltrust.org.uk
Eskdale Mill, Boot, Eskdale, Holmrook CA19 1TG.
Telephone: 01946 723335. Website: www.eskdalemill.co.uk
Gleaston Watermill, Gleaston, near Ulverston LA12 0QH.
Telephone: 01229 869244. Website: www.watermill.co.uk
Heron Corn Mill, Waterhouse Mills, Beetham, Milnthorpe LA7 7AR. Telephone: 01539 565027. Website: www.heronmill.org
Little Salkeld Watermill, Little Salkeld, Penrith CA10 1NN.
Telephone: 01768 881523. Website: www.organicmill.co.uk

DERBYSHIRE

Caudwell's Mill, Bakewell Road, Rowsley, Matlock DE4 2EB.
Telephone: 01629 734374. Website: www.caudwellscrafts.co.uk
Stainsby Mill, Hardwick Estate, Doe Lea, Chesterfield S44 5QJ.
Telephone: 01246 850430. Website: www.nationaltrust.org.uk

DEVON

Clyston Mill, Killerton, Broadclyst, Exeter EX5 3EW.
Telephone: 01392 462425. Website: www.nationaltrust.org.uk
Manor Mill, Branscombe, Seaton EX12 3DB. Telephone: 01392 881691.
Website: www.nationaltrust.org.uk
Otterton Mill, Otterton, Budleigh Salterton EX9 7HG.
Telephone: 01395 568521. Website: www.ottertonmill.com

DORSET

Mangerton Mill, Mangerton, Bridport DT6 3SG. Telephone: 01308 485224.
Sturminster Mill, Sturminster Newton. Telephone: 01747 854355.
Website: www.sturminsternewton-museum.co.uk
The Town Mill, Mill Lane, Lyme Regis DT7 3PU. Telephone: 01297 443579.
Website: www.townmill.org.uk
White Mill, Sturminster Marshall, near Wimborne Minster BH21 4BX.
Telephone: 01258 858051. Website: www.nationaltrust.org.uk/whitemill, also www.whitemill.org

COUNTY DURHAM

Path Head Watermill, Summerhill, Blaydon, Tyne and Wear NE21 4SP.
Telephone: 0191 414 6288. Website: www.gatesheadmill.co.uk

ESSEX

Bourne Mill, Bourne Road, Colchester CO2 8RT.
   Telephone: 01206 572422. Website: www.nationaltrust.org.uk
Thorrington Tide Mill, Thorrington, near Brightlingsea.
   Telephone: 01245 437663. Website: www.essexcc.gov.uk

HAMPSHIRE

Alderholt Mill, Sandleheath Road, Alderholt, Fordingbridge SP6 1PU.
   Telephone: 01425 653130. Website: www.alderholtmill.co.uk
City Mill, Bridge Street, Winchester SO23 0EJ. Telephone: 01962 870057.
   Website: www.nationaltrust.org.uk
Eling Tide Mill, The Tollbridge, Totton, Southampton SO40 9HF.
   Telephone: 023 8086 9575. Website: www.elingtidemill.wanadoo.co.uk

HEREFORDSHIRE

Mortimer's Cross Mill, Lucton, Leominster HR6 9PE.
   Telephone: 01568 708820. Website: www.midlandmillsopen.org.uk

HERTFORDSHIRE

Mill Green Mill, Mill Green, Hatfield AL9 5PD. Telephone: 01707 271362.
   Website: www.welhat.gov.uk
Redbournbury Mill, Redbournbury Lane, Redbourn Road, St Albans AL3 6RS.
   Telephone: 01582 792874. Website: www.redbournmill.co.uk

ISLE OF WIGHT

Calbourne Watermill and Museum, Newport Road, Calbourne PO30 4JN.
   Telephone: 01983 531227. Website: www.calbournewatermill.co.uk

KENT

Crabble Corn Mill, Lower Road, River, Dover CT17 0UY.
   Telephone: 08701 453857. Website: www.ccmt.org.uk

LINCOLNSHIRE

Alvingham Mill, Church Lane, Alvingham. Telephone: 01507 327544.
Cogglesford Mill, East Road, Sleaford. Telephone: 01529 414294.
   Website: www.tic.oden.co.uk

LONDON

House Mill, Three Mills Lane, Bromley-by-Bow, London E3 3DU.
   Telephone: 020 8980 4626.

MILLS TO VISIT

NORFOLK

Letheringsett Mill, Riverside Road, Letheringsett, Holt NR25 7YD.
   Telephone: 01263 713153. Website: www.letheringsettwatermill.co.uk

NORTHUMBERLAND

Heatherslaw Mill, Ford, Cornhill-on-Tweed. Telephone: 01890 820488.
   Website: www.fordetal.co.uk

NOTTINGHAMSHIRE

Ollerton Mill, Market Place, Ollerton, Newark NG22 9AA.
   Telephone: 01623 822469 or 824094.

OXFORDSHIRE

Coleshill Mill, Coleshill, Swindon. Telephone: 01793 762209.
   Website: www.nationaltrust.org.uk
Mapledurham Mill, Mapledurham, Reading RG4 7TR.
   Telephone: 0118 972 3350. Website: www.mapledurham.co.uk

SHROPSHIRE

Daniel's Mill, Eardington, Bridgnorth WV16 5JL. Telephone: 01746 762753.

SOMERSET

Bishop's Lydeard Mill, Mill Lane, Bishop's Lydeard, Taunton TA4 3LN.
   Telephone: 01823 432151.
Burcott Mill, Wookey, Wells BA5 1NJ. Telephone: 01749 673118.
   Website: www.burcottmill.com
Dunster Working Watermill, Mill Lane, Dunster TA24 6SW.
   Telephone: 01643 821759. Website: www.nationaltrust.org.uk
Gants Mill, Bruton BA10 0DB. Telephone: 01749 812393.
   Website: www.gantsmill.co.uk

STAFFORDSHIRE

Brindley Mill, Mill Street, Leek ST13 8HA. Telephone: 01538 483741.
   Website: www.brindleymill.net
Shugborough Estate Mill, Milford, near Stafford ST17 0XB.
   Telephone: 01889 881388. Website: www.shugborough.org.uk

SUFFOLK

Alton Mill, Museum of East Anglian Life, Stowmarket IP14 1DL.
   Telephone: 01449 612229. Website: www.eastanglianlife.org.uk
Pakenham Watermill, Mill Road, Pakenham, Bury St Edmunds IP31 2NB.
   Telephone: 01284 724075.

Woodbridge Tide Mill, Tide Mill Way, Woodbridge IP12 1AP.
Telephone: 01473 626618.
Website: http://woodbridgesuffolk.info/WB-Attractions/TideMill/

SURREY

Cobham Mill, Mill Road, Cobham KT11 3AL. Telephone: 01932 867387.
Website: www.cobhamheritage.org.uk
Shalford Mill, Shalford, near Guildford GU4 8BS.
Telephone: 01483 561389. Website: www.nationaltrust.org.uk

SUSSEX

Lurgashall Mill, Weald and Downland Open Air Museum, Singleton, Chichester PO18 0EU. Telephone: 01243 811363. Website: www.wealddown.co.uk
Michelham Priory Mill, Upper Dicker, near Hailsham BN27 3QS.
Telephone: 01323 844224. Website: www.sussexpast.co.uk
Park Mill, Bateman's, Burwash, Etchingham TN19 7DS.
Telephone: 01435 882302. Website: www.nationaltrust.org.uk

WARWICKSHIRE

Charlecote Mill, Hampton Lucy CV35 8BB. Telephone: 01789 842072.
Website: www.charlecotemill.co.uk
New Hall Mill, Wylde Green Road, Sutton Coldfield.
Telephone: 0121 355 3265. Website: www.newhallmill.org.uk
Sarehole Mill, Cole Bank Road, Hall Green, Birmingham B13 0BD.
Telephone: 0121 777 6612. Website: www.bmag.org.uk
Wellesbourne Mill, Kineton Road, Wellesbourne CV35 9HG.
Telephone: 01789 470470. Website: www.wellesbournemill.co.uk

YORKSHIRE

Fountains Abbey Mill, Fountains Abbey and Studley Royal Estate, Fountains, Ripon HG4 3DY. Telephone: 01765 608888.
Website: www.fountainsabbey.org.uk
Raindale Mill, York Castle Museum, Eye of York, York YO1 9PY.
Telephone: 01904 653611. Website: www.yorkcastlemuseum.org.uk
Tockett's Mill, Skelton Road, Guisborough.
Website: www.redcar-cleveland.gov.uk
Worsbrough Mill Museum, Worsbrough Bridge, Barnsley S70 5LJ.
Telephone: 01226 774527. Website: www.barnsley.gov.uk

CHANNEL ISLANDS

Quetivel Mill, St Peter's Valley, Jersey. Telephone: 01534 483193.
Website: www.nationaltrustjersey.org.je

MILLS TO VISIT

NORTHERN IRELAND

Annalong Mill, Marine Park, Annalong, County Down BT34 4RH.
  Telephone: 028 4175 2256. Website: www.visitnewryandmourne.com
Castle Ward Mills, Downpatrick Road, Strangford, County Down BT30 7LS.
  Telephone: 028 4488 1204. Website: www.nationaltrust.org.uk

SCOTLAND

Barony Mills, Birsay, Orkney KW17 2LY. Telephone: 01856 721439.
  Website: www.birsay.org.uk/baronymill.htm
Barry Mill, Barry, Carnoustie, Angus DD7 7RJ. Telephone: 01241 856761.
  Website: www.nts.org.uk
Blair Atholl Mill, Blair Atholl, Perthshire PH18 5SH.
  Telephone: 01796 481321. Website: www.blairathollwatermill.co.uk
Click Mill, Dounby, Orkney. Telephone: 01856 841815.
Crofthouse Museum, South Voe, Dunrossness, Shetland.
  Telephone: 01595 695057. Website: www.shetland-museum.org.uk/crofthouse
New Abbey Corn Mill, New Abbey, Dumfries DG2 8BX.
  Telephone: 01387 850260. Website: www.historic-scotland.gov.uk
Preston Mill, East Linton, East Lothian. Telephone: 01620 860426.
Quendale Mill, Dunrossness, Shetland ZE2 9JD. Telephone: 01950 460969.
Website: www.quendalemill.shetland.co.uk
Shawbost Mill, Shawbost, Lewis, Western Isles. Telephone: 01851 710208.

WALES

Bacheldre Watermill, Church Stoke, Montgomery, Powys SY15 6TE.
  Telephone: 01588 620489. Website: www.bacheldremill.co.uk
Carew Tidal Mill, Carew, Tenby, Pembrokeshire SA70 8SL.
  Telephone: 01646 651782. Website: www.carewcastle.com
Y Felin, Mill Street, St Dogmaels, Cardigan SA43 3DY.
  Telephone: 01239 613999. Website: www.welshmills.org.uk
Y Felin Dolws, Gower Heritage Centre, Parkmill, Swansea SA3 2EH.
  Telephone: 01792 371206. Website: www.gowerheritagecentre.co.uk
Gelligroes Mill, Pontllanfraith, Blackwood NP12 2HY.
  Telephone: 01495 222322.
Melin Bompren, St Fagans National History Museum, St Fagans, Cardiff CF5
  6XB. Telephone: 029 2057 3500. Website: www.museumwales.ac.uk
Melin Howell, Llanddeusant, Anglesey. Telephone: 01407 730240.

WINDMILLS

(P) = post mill; (S) = smock mill; (T) = tower mill.

BEDFORDSHIRE

Stevington Mill (P), Stevington, Bedford. Telephone: 01234 824330.
   Website: www.bedfordshire.gov.uk

BUCKINGHAMSHIRE

Lacey Green Mill (S), Lacey Green, near Princes Risborough.
   Telephone: 01844 345360. Website: www.chilternsociety.org.uk
Pitstone Mill (P), Ivinghoe. Telephone: 01442 851227.
   Website: www.nationaltrust.org.uk
Quainton Mill (T), Quainton, Aylesbury. Telephone: 01296 655306.
   Website: www.quainton.net

CAMBRIDGESHIRE

Bourn Mill (P), Bourn, Cambridge. Telephone: 01223 243830.
   Website: www.cpswandlebury.org
Cattell's Mill (S), Mill Road, Willingham. Telephone: 01954 261168.
Chishill Mill (P), Great Chishill. Telephone: 01223 718131.
Foster's Mill (T), Swaffham Prior CB5 0JZ. Telephone: 01638 741009.
   Website: www.fostersmill.co.uk
Fulbourn Mill (S), Fulbourn. Telephone: 01223 880649.
   Website: www.fulbourn.windmill.btinternet.co.uk
Great Gransden Mill (P), Mill Road, Great Gransden.
   Telephone: 01767 677487.
Impington Mill (S), Cambridge Road, Impington, Cambridge.
   Telephone: 01223 232284.
Madingley Mill (P), St Neots Road, Madingley, Cambridge.
   Telephone: 01954 211047.
Over Mill (T), Over. Telephone: 01954 230742.
Stevens Mill (T), Mill Lane, Burwell CB25 0HL. Telephone: 01638 742847.
   Website: http://mysite.freeserve.com/burwell_museum
Wicken Mill (S), Wicken. Telephone: 01664 822751.
   Website: www.geocities.com/wickenmill/

CHESHIRE

Bidston Mill (T), Bidston Hill, Birkenhead. Telephone: 0151 653 9332.

DERBYSHIRE

Heage Mill (T), Belper. Telephone: 01773 853579.
   Website: www.heagewindmill.co.uk

MILLS TO VISIT

COUNTY DURHAM

Fulwell Mill (T), Newcastle Road, Fulwell, Sunderland SR5 1EX.
   Telephone: 0191 516 9790. Website: www.fulwell-windmill.com

ESSEX

Aythorpe Roding Mill (P), near Leaden Roding. Telephone: 01245 437663.
   Website: www.essexcc.gov.uk
Bocking Mill (P), off Church Street, Bocking. Telephone: 01376 324781.
Finchingfield Mill (P), Haverhill Road, Finchingfield.
   Telephone: 01245 437663. Website: www.essexcc.gov.uk
John Webb's Mill (T), Thaxted, Dunmow. Telephone: 01371 830285.
Mountnessing Mill (P), near Brentwood. Telephone: 01245 437663.
   Website: www.essexcc.gov.uk
Stansted Mountfitchet Mill (T), Mill Hill, Stansted Mountfitchet CM24 8XX.
   Telephone: 01279 647213.
   Website: www.stanstedmountfitchetwindmill.co.uk
Stock Mill (T), Mill Lane, Stock, Chelmsford. Telephone: 01245 437663.
   Website: www.essexcc.gov.uk
Upminster Mill (S), Upminster. Telephone: 01708 226040.
   Website: www.upminsterwindmill.co.uk

HAMPSHIRE

Bursledon Mill (T), Windmill Lane, Bursledon, Southampton SO31 8BG.
   Telephone: 0845 603 5635. Website: www.hants.gov.uk/museum/windmill

HERTFORDSHIRE

Cromer Mill (P), Cromer, near Stevenage. Telephone: 01279 843301.
   Website: www. hertsmuseums.org.uk/cromer-windmill

ISLE OF WIGHT

Bembridge Mill (T), High Street, Bembridge PO35 5SQ.
   Telephone: 01983 873945. Website: www.nationaltrust.org.uk

KENT

Draper's Mill (S), St Peter's Footpath, off College Road, Margate CT9 2SP.
   Telephone: 01843 226227.
Herne Mill (S), Mill View Road, Herne CT6 7DR.
   Telephone: 01227 361326. Website: http://kentwindmills.homestead.com
Meopham Mill (S), Meopham Green. Telephone: 01474 813518.
Sarre Mill (S), Ramsgate Road, Sarre, Isle of Thanet CT7 0JU.
   Telephone: 01843 847573. Website: www.sarremill.co.uk
Stelling Minnis Mill (S), Mill Lane, Stelling Minnis, Canterbury.

Telephone: 01227 709550.
Website: www.stelling-minnis.co.uk/windmill
Stock's Mill (P), Rye Road, Wittersham, Tenterden TN30 7ER.
Telephone: 01797 270295.
Union Mill (S), The Hill, Cranbrook. Telephone: 01580 712984.
Website: www.unionmill.org.uk
White Mill (S), The Causeway, Ash Road, Sandwich CT13 9JB.
Telephone: 01304 612076.
Website: http://whitemill.open-sandwich.co.uk
Willesborough Mill (S), Mill Lane, Willesborough, Ashford TN24 0GQ.
Telephone: 01233 661866. Website: www.willesboroughwindmill.co.uk
Woodchurch Mill (S), Tenterden. Telephone: 01233 860649.
Website: www.woodchurchwindmill.co.uk

LANCASHIRE

Lytham Mill (T), Lytham Green, Lytham FY8 5LD.
Telephone: 01253 794879. Website: www.lythamheritage.fsnet.co.uk
Marsh Mill (T), 76 Barton Street, Thornton Cleveleys FY5 4AE.
Telephone: 01253 860765.

LEICESTERSHIRE

Hough Mill (T), off St George's Hill, Swannington.
Telephone: 01530 832704. Website: www.swannington-heritage.co.uk
Wymondham Mill (T), Butt Lane, Wymondham, near Melton Mowbray LE14 2BU. Telephone: 01572 787304.
Website: www.wymondhamwindmill.co.uk

LINCOLNSHIRE

Alford Mill (T), East Street, Alford LN13 9EQ. Telephone: 01507 462136. Website: www.alfordtown.co.uk/shared/mill/mill.htm
Dobson's Mill (T), High Street, Burgh le Marsh. Telephone: 01754 766658. Website: http://burghlemarsh.info/town/windmill/windmill.htm
Ellis's Mill (T), Mill Road, Lincoln LN1 3LY. Telephone: 01522 528448.
Heckington Mill (T), Heckington, Sleaford. Telephone: 01529 461919.
Hewitt's Windmill (T), Heapham, Gainsborough. Telephone: 01427 838230.
Maud Foster Mill (T), Willoughby Road, Boston PE21 9EG.
Telephone: 01205 352188. Website: www.maudfoster.co.uk

The fine Maud Foster Mill, at Boston, Lincolnshire. Built by the Hull millwrights Norman & Smithson in 1819, the mill was restored by the Waterfield family in the 1980s and is still at work by wind, producing a range of stoneground flours.

Moulton Mill (T), Moulton, Spalding PE12 6QD.
   Telephone: 01406 373368.
Mount Pleasant Mill (T), North Cliff Road, Kirton-in-Lindsey DN21 4NH.
   Telephone: 01652 640177.
   Website: www.mountpleasantwindmill.co.uk
Trader Mill (T), Frithville Road, Sibsey, Boston PE22 0SY.
   Telephone: 01205 750036. Website: www.sibsey.fsnet.co.uk
Waltham Mill (T), Brigsley Road, Waltham, Grimsby DN37 0JZ.
   Telephone: 01472 752122. Website: www.walthamwindmill.co.uk
Wrawby Mill (P), Mill Lane, Wrawby, Brigg. Telephone: 01652 653699.

LONDON

Wimbledon Mill (S), Wimbledon Common, London SW19 5NR.
   Telephone: 020 8947 2825.
   Website: www.wimbledonwindmillmuseum.org.uk

NORFOLK

Billingford Mill (T), Billingford Common, Billingford, Diss IP21 4HL.
   Telephone: 01603 222705.
Bircham Windmill (T), Great Bircham, King's Lynn PE31 6SJ.
   Telephone: 01485 578393. Website: www.birchamwindmill.co.uk
Denver Mill (T), Denver, Downham Market. Telephone: 01366 384009.
   Website: www.denvermill.co.uk
Garboldisham Mill (P), Garboldisham. Telephone: 01953 681593.
Little Cressingham Wind and Watermill (T), Watton. Telephone: 01603 222705.
Old Buckenham Mill (T), Green Lane, Old Buckenham, Attleborough NR17 1AA. Telephone: 01903 454371 or 01603 222705.
Sutton Mill (T), Windmill Lane, Sutton NR12 9RZ.
   Telephone: 01692 581195. Website: www.suttonwindmill.co.uk
Wicklewood Mill (T), Wymondham. Telephone: 01603 222705.

NOTTINGHAMSHIRE

Green's Mill (T), Windmill Lane, Sneinton, Nottingham NG2 4QB.
   Telephone: 0115 915 6878. Website: www.greensmill.org.uk
North Leverton Mill (T), Mill Lane, North Leverton, Retford DN22 0AB.
   Telephone: 01427 880573.

OXFORDSHIRE

Chinnor Mill (P), Mill Lane, Chinnor. Telephone: 01844 292095.
   Website: www.oxfordshire.gov.uk
Wheatley Mill (T), Windmill Lane, Wheatley. Telephone: 01865 874610.
   Website: http://www.advsys.co.uk/wheatleymill

RUTLAND

Whissendine Mill (T), Melton Road, Whissendine, Oakham LE15 7EU.
  Telephone: 01664 474172.

SHROPSHIRE

Asterley Mill (T), Minsterley. Telephone: 01743 791434.

SOMERSET

Ashton Mill (T), Chapel Allerton, Wedmore. Telephone: 01934 712034.
  Website: www.sedgemoor.gov.uk/sedgemoorweb/museum.htm
Stembridge Mill (T), High Ham TA10 9DJ. Telephone: 01458 250818.
  Website: www.nationaltrust.org.uk

STAFFORDSHIRE

Broad Eye Mill (T), Castle Hill, Stafford. Telephone: 01785 611737.
  Website: www.broadeyewindmill.co.uk

SUFFOLK

Bardwell Mill (T), Bardwell, Ixworth. Telephone: 01359 251331.
Buttrum's Mill (T), Burkitt Road, Woodbridge IP12 4JJ.
  Telephone: 01473 264755.
Drinkstone Mills (P and S), Woolpit Road, Drinkstone, near Bury St Edmunds
  IP30 9SP. Telephone: 07843 074700.
Pakenham Windmill (T), Thieves Lane, Pakenham, Bury St Edmunds IP31 2NF.
  Telephone: 01359 270570.
Saxtead Green Mill (P), Framlingham. Telephone: 01728 685789 or
  01760 755161.
Stanton Windmill (P), Mill Farm, Upthorpe Road, Stanton, near Bury St
  Edmunds IP31 2AW. Telephone: 01359 250622.
Thelnetham Mill (T), Mill Road, Thelnetham IP22 1JZ.
  Telephone: 01473 727853.

SURREY

Lowfield Heath Mill (P), Rusper Road, Charlwoood.
  Telephone: 01293 862374 or 409845.
Outwood Mill (P), Outwood, Redhill RH1 5PW. Telephone: 07760 194948.
  Website: www.outwoodwindmill.co.uk
Shirley Mill (T), Post Mill Close, Shirley, Croydon CR0 5DY.
  Telephone: 020 8406 4676. Website: www.croydononline.org

SUSSEX

Argos Hill Mill (P), Mayfield. Telephone: 01453 873367.

High Salvington Mill (P), Furze Road, High Salvington, Worthing BN13 3BP.
   Telephone: 01903 262443. Website: www.highsalvingtonwindmill.co.uk
Jill Mill (P), Clayton. Telephone: 01273 843263.
   Website: www.jillwindmill.org.uk
Nutley Mill (P), Crowborough Road, Nutley, Crowborough TN20 6UP.
   Telephone: 01453 873367. Website: www.udps.co.uk
Oldland Mill (P), Keymer. Telephone: 01273 503747.
   Website: www.oldlandwindmill.co.uk
Polegate Mill (T), Park Croft, Polegate BN26 5LB.
   Telephone: 01323 644727. Website: www.sussexmillsgroup.org.uk
Shipley Mill (S), near Horsham. Telephone: 01403 730439.
   Website: www.shipleywindmill.org.uk
Stone Cross Mill (T), Stone Cross. Telephone: 01323 763206 or 763253.
West Blatchington Mill (S), Holmes Avenue, Hove. Telephone: 01273 776017.

WARWICKSHIRE

Chesterton Mill (T), Chesterton, Warwick. Telephone: 01926 412033.
Tysoe Mill (T), Compton Wynyates. Telephone: 01295 680229.

WILTSHIRE

Wilton Mill (T), Wilton, Marlborough SN8 3LT. Telephone: 01672 870202.
   Website: www.wiltonwindmill.co.uk

WORCESTERSHIRE

Danzey Green Mill (P), Avoncroft Museum of Historic Buildings, Stoke Heath,
   Bromsgrove B60 4JR. Telephone: 01527 831363 or 831886.
   Website: www.avoncroft.org.uk

YORKSHIRE

Skidby Mill (T), Skidby, Cottingham HU16 5TF. Telephone: 01482 392773 or
   392777.

NORTHERN IRELAND

Ballycopeland Mill (T), Windmill Road, Millisle, Newtownards, County Down
   BT22 2DS. Telephone: 028 9054 6552.
   Website: www.ehsni.gov.uk/places/monuments/ballycopeland.shtml

WALES

Melin Llynnon (T), Llanddeusant, Anglesey. Telephone: 01407 730797.
   Website: www.ynysmon.gov.uk

INDEX

Abu Hureyra 9
Africa 10
Albion Mill 15, 15
Anglesey 31
Anglo-Saxon 12, 33
Animal feed 5, 7, 18, 19, 33, 31, 34
Animal power 11, 11, 40
Ark, meal 39, 40, 44, 44
Arrow Mill, Kingsland 4, 41, 45, 50
Asia 9
Automatic mill 15, 16
Baker, bakery 6, 11, 11, 40, 48
Balancing, millstones 34, 35
Barley 6–7, 6, 9, 26, 27
Barry 19
Bedstone 9, 10, 11, 32, 33, 34
 (see also lower stone)
Beehive quern 11
Bill, mill 35
Bin 14, 21–2, 23, 39
Blezzard, Dick 46
Bolter 14, 39, 40, 41, 41, 42, 50
Bossava Mill, Paul 40
Bran 4, 16, 17, 31, 32, 36, 37, 39, 40, 41, 42, 42
Brantham Mill 22
Bread 5, 6, 7, 15, 19, 31, 39, 50
Brewing 35
Briddon and Fowler 37
Bristol Cathedral 48
Britain 5, 6, 7, 9, 11, 12, 16, 17, 18, 26, 31, 32, 49, 50
Bronze Age 7, 9
Buchholz, Gustav 17
Bühler 37
Bunbury Mill 28
Burr, French 31–2, 33, 34, 50
 (see also French millstones)
Carew Mill 26
Cartwright, William 50
Caudwell's Mill 37, 41
Centrifugal 9, 43, 43
Chaucer, Geoffrey 47
Chedworth 12
Chester 17
Clarke and Dunham 34
Cleaner, grain 18, 24, 26
Cleaning grain 25–6, 50
Cockle cylinder 25–6, 25, 27 (see also Trieur)
Coleshill Mill 14
Composition millstone 30, 32–3, 34
Conveyor 15, 16, 22
Corn Laws 15
Crowdy Mill, Harberton 23, 33
Croydon 17
Cullin stone 31, 33
Dell, William 9
Domesday Book 13, 47
Donkey mill 11, 11
Dressing, flour 18, 19, 25, 31, 36–7, 39–43, 48, 50
Dressing, millstones 9, 12, 32, 32–5, 34
Drying, grain 26–9
Dry stoner 26, 27
Ebbsfleet 12
Egypt 6, 9
Elevator 15, 16, 21, 23

Elsecar Mill 41
Emmer 6
Endosperm 4, 36, 37
England 7, 12, 13, 16,17, 21, 28, 31, 42, 47, 48
Ergot 7
Europe 7, 9, 16
Evans, Oliver 15, 16, 21
Eye 9, 11, 32, 33, 34
Fan 23, 25
Fantail 15, 15
Felin Newydd, Crug-y-Bar 42
Fison and Co 17
France 16, 25, 31, 42
French millstones 31–2, 33, 34, 35, 50
Frost and Son 17
Furrow 30, 32–4, 33, 35
Gardner and Son 18
Gearing 12, 12, 13, 14, 14, 40, 49, 50
German millstones 31 (see also Cullin stone)
Gluten 6, 7
Gradual reduction 17, 17, 25, 36, 39, 43, 50
Granite millstone 31, 32
Great Alne Mill 43
Grist mill 13, 48
Groats 35
Hadrian's Wall 12
Hampshire 12
Handle 10, 11
Handley, R. G. 33
Heage Mill 15
Heron Mill 34, 38, 39
High milling 16
Hoist 14, 20, 21–2, 23
Hopkinson 43
Horizontal waterwheel 12, 12, 13
Houghton Mill 34, 35, 36
Howes, Howes and Ewell 24, 27
Hughes, J. and Sons 33
Hulling 6, 17
Hungary 16, 17
Hyde Mill, Ickleford 22
Ickleford Mill 19
India 26
Ireland 47, 49
Iron Age 9
Iron rolls 7, 16, 17
Islington 18
Italy 11, 11
Jog-scry 38, 39
Kent 12
Kiln 7, 26–7, 28–9, 28, 29
Knettishall, Diss 5
Lands 30, 32–3, 33, 33, 35
Lapford Mill 23
Lee, Edmund 15
Lightening 11
Liverpool 17
London 8, 11, 15, 15, 18, 25, 30, 33
Longhill Mill, Urquhart 7
Lower stone 9, 10
Low milling 16
Lyng 22
Malt 7, 16, 31, 35
Manchester 17
Manor Mill, Branscombe 44

Mark Lane 17, 24, 25, 27
Maslin 7
Maud Foster Mill, Boston 43, 60
Meal 7, 9, 10, 11, 16, 19, 21, 31, 35, 39–40, 42, 44, 47
Medieval 7, 13, 13, 14, 21, 25, 39, 40, 47–8, 48, 49
Mediterranean 12
Merchant mill 15, 48
Middle Ages 7, 13, 14, 21, 32, 48, 49 (see also Medieval)
Miller 6, 13, 14, 15, 16, 17, 21, 22, 28, 33, 45, 47–8, 48
Millers' Mutual Association 18
Mill of Mundurno 23, 28
Millstone Grit 31
Millstones 5, 7, 7, 9, 12, 12, 13, 13, 14, 14, 15, 16, 17, 18, 19, 21, 30–5, 31–5, 38, 40, 44
Millwright 17, 18, 21, 33, 45, 46, 49–50
Milne, John 39, 40
Moisture content 26, 27, 28
Montgarrie Mill 28
Mortar 9
Mouture économique 16
Mow Cop 31
Muncaster Mill 29
NABIM 17–18
Nafford 18
Neolithic 9
Nuneaton 14
Oatmeal mill 19, 21, 23, 26, 27, 28, 38, 39
Oats 7, 7, 20, 26, 27, 29, 35
Offley Mill 19
Ollerton Mill 49
Organic 19
Osney Mill 20, 21
Overshot waterwheel 13
Peak millstones 31, 32, 34
Pearl/pot barley 7, 7
Plan sifter 39, 41
Pompeii 11, 11
Porcelain rolls 17, 17, 36
Port mill 15, 19, 26
Post mill 13, 13, 14, 15, 33
Purifier 8, 37
Quern 9, 10, 11
Radford and Sons 17
Rank, Joseph 19
Reel 9, 39, 42–3, 43
Robinson 19, 37
Roller mill 5, 7, 8, 16–17, 17, 18, 18, 19, 33, 35, 36–7, 36–7, 42, 50
Roman 11, 12, 31, 33, 39
Rotary cleaner 25
Rotary quern 9, 10, 11
Rubble reel 26
Runner stone 9, 34, 35, 35
Russia 19, 26
Rye 7, 7, 31, 50
Rynd 11, 32, 33
Sack barrow 21
Sacks 14, 21–2, 23, 44–5, 48
Saddle quern 9, 10
Sails 13, 13, 15, 15, 47
San Miguel, Salvador 37
Scalper 8, 36–7
Scotland 7, 27, 28

Seck Brothers 17
Semolina 16, 17, 36
Shaw Mill 34
Shelling 27, 29, 34, 35, 39
Shetland mill 12
Sieve 16, 23, 25, 26, 35, 37, 39–43, 40
Silex Works 30
Simmons and Morten's Mill 8
Simon, Henry 17, 36, 43
Smock mill 13
Smutter 24, 25, 26
Soke 47, 48
Spelt 6
Spindle 10, 11, 12, 32, 35
Stanton Mill 33
Steam mill, power 9, 15, 19, 22, 48
Storage 13, 21, 27
Sulzberger, Jacob 16–17
Switzerland 16–17
Syria 9
Tadcaster 17
Tamworth 12
Tentering 11
Thornton Abbey 34
Threshing 6, 25
Tide mill 12
Timms Mill, Goole 27, 37
Toll 39, 44, 47–8
Tower mill 13, 13, 15, 15, 46
Traditional Corn Millers' Guild 19
Trafford Mill 35
Transport 21–2, 22–3, 32
Treble mill 14, 14
Trewey Mill 32
Trieur 25–6, 27
Tuttle Hill Mill 46
Undershot waterwheel 12
United States of America 15, 17, 19, 25
Upper Kennerty Mill 27
Upper stone 9, 10, 11 (see also Runner stone)
Ventilation, millstone 35, 35
Vertical waterwheel 12, 12, 13, 13
Vienna 18
Viver Mill, Hincaster 29
Wagon 20, 21, 22
Wales 28, 31, 45
Walworth kiln 29
Walzmühle, Budapest 3, 17
Washing, grain 26
Water power 11–13, 18, 19, 37, 49, 50
Wegmann 17
Weighing 44, 45
Welsh millstones 31
Westminster Bridge 8
Wheat 4, 6, 6, 7, 7, 9, 16, 19, 20, 23, 26, 29, 31, 45
White flour 16, 18, 26, 31, 32, 36, 39, 43
White Mill 26
Wind power 13, 15, 19, 40, 49, 50
Winnowing 25, 45
Wire machine 39, 42, 42
Worsbrough Mill 40
Zones, milling 9, 10, 12, 32